T0231679

Microoptics
and Nanooptics
FABRICATION

Microoptics
and Nanooptics
FABRICATION

Edited by
Shanalyn A. Kemme

CRC Press
Taylor & Francis Group
Boca Raton   London   New York

CRC Press is an imprint of the
Taylor & Francis Group, an **informa** business

CRC Press
Taylor & Francis Group
6000 Broken Sound Parkway NW, Suite 300
Boca Raton, FL 33487-2742

First issued in paperback 2017

© 2010 by Taylor and Francis Group, LLC
CRC Press is an imprint of Taylor & Francis Group, an Informa business

No claim to original U.S. Government works

ISBN 13: 978-1-138-11644-3 (pbk)
ISBN 13: 978-0-8493-3676-8 (hbk)

Library of Congress Cataloging-in-Publication Data

Microoptics and nanooptics fabrication / editor, Shanalyn Kemme.
    p. cm.
  "A CRC title."
  Includes bibliographical references and index.
  ISBN 978-0-8493-3676-8 (hardcover : alk. paper)
  1. Nanostructured materials--Design and construction. 2. Microfabrication. 3. Nanophotonics. 4. Nanostructures--Optical properties. I. Kemme, Shanalyn. II. Title.

TA418.9.N35M533 2010
681'.4--dc22
2009027804

Visit the Taylor & Francis Web site at
http://www.taylorandfrancis.com

and the CRC Press Web site at
http://www.crcpress.com

This book is dedicated to the memory of James G. Fleming. Many of us were lucky enough to have collaborated with Jim. He was reserved and unassuming, but we all know that he was the brains and hands behind much of the foundational semiconductor optical fabrication such as his three-dimensional photonic crystal work (J.G. Fleming, et al. Nature, *vol. 417, pp. 52–55).*

Contents

Preface

The initial popularity of microoptics and nanooptics is easy to understand when considering a couple of points: small, integrated optical components are thermally and mechanically more robust than their larger counterparts, and microoptical components match the small scale of commonly employed sources and detectors. While these driving reasons are certainly still valid, even more exciting advances continue to fuel the growth of micro- and nanooptics. Fabrication at such a small, precise scale makes possible new optical components that provide access to the physical optics regime, even down to the ultraviolet wavelength band, affecting properties such as polarization through subwavelength feature definition.

This last point underscores the vital relationship between the concept of a micro/nanooptical component and its associated fabrication technology. In practice, micro- and nanooptical-component-fabrication technologies are so intimately correlated with micro- and nanooptical-component design that the two tasks cannot be decoupled.

Unfortunately, fabrication of microoptical and nanooptical components is practically constrained by material selection, component lateral extent, and/or minimum feature size. This is why microoptical and nanooptical-fabrication techniques often lag behind the corresponding theoretical component development. While we are unbound in considering a component with optimal refractive indices, with precisely the required shape and positioned perfectly within the optical path, it is the experimental realization of this component that often determines success (or failure).

A micro- or nanooptical-fabrication approach is so closely connected to the resulting optical component that a change in the fabrication approach can spawn a new optical component. Consider the revolutionary field of photonic crystals. Current literature abounds with one-dimensional and two-dimensional photonic crystal examples that the optics community would previously have referred to as thick films, corrugations, or gratings. However, the development of new two-dimensional and three-dimensional photonic crystal fabrication approaches lifted many material, extent, and precision limitations and allowed component configurations not previously possible.

By contrast, the broader field of diffractive optical element fabrication is virtually established, compared to photonic crystal fabrication technology. This is largely due to a diffractive optics fabrication approach, called binary optics (Swanson and Veldkamp) that takes full advantage of the mature semiconductor fabrication process. Even with this leverage, every step in the process must be reformulated to achieve the necessary optical targets, from choice of materials, to mask definition, to etch/deposition steps that produce surfaces of specified flatness and finish, rarely requested in the field of semiconductor fabrication.

As diffractive optical element feature sizes decrease and aspect ratios increase, successful diffractive optics components will continue to expand from the microwave to the infrared through the visible regime. And as the fidelity of replication processes improves, diffractive optical components will creep into the commercial world (accompanied by the inevitable reduction in cost).

Because of the connection between micro/nanooptical component concept and fabrication illustrated earlier, this book explores in detail successful fabrication processes associated with micro- and nanooptical components. We document the state of the art in fabrication processes as they directly affect a micro- or nanooptical component's intended performance. This book is written keeping the professional optical engineer in mind, focusing on key tricks of the trade rather than broad, well-published processing fundamentals. Each contributing author is an expert in the field of fabrication technology, and as an ensemble represents the vanguard in microoptical fabrication today. Ever-changing technologies, like those utilized in micro- and nanooptical fabrication, call for up-to-date synopses. I eagerly grabbed the opportunity presented to me by Taylor & Francis Books and these contributing authors to help this one come about.

MATLAB® is a registered trademark of The MathWorks, Inc. For product information, please contact:

The MathWorks, Inc.
3 Apple Hill Drive
Natick, MA 01760-2098 USA
Tel: 508 647 7000
Fax: 508-647-7001
E-mail: info@mathworks.com
Web: www.mathworks.com

Contributors

Gregg T. Borek
MEMS Optical, Inc.
Huntsville, Alabama

Caihua Chen
Department of Electrical and
    Computer Engineering
University of Delaware
Newark, Delaware

Alvaro A. Cruz-Cabrera
Photonic Microsystems
    Technologies
Sandia National Laboratories
Albuquerque, New Mexico

Ihab F. El-Kady
Photonic Microsystems
    Technologies
Sandia National Laboratories
Albuquerque, New Mexico

Aaron Gin
Photonic Microsystems Technologies
    and the Center for Integrated
    Nanotechnologies (CINT)
Sandia National Laboratories
Albuquerque, New Mexico

Eric G. Johnson
The Center for Optoelectronics and
    Optical Communications
University of North Carolina at
    Charlotte
Charlotte, North Carolina

Shanalyn A. Kemme
Photonic Microsystems
    Technologies
Sandia National Laboratories
Albuquerque, New Mexico

Janusz Murakowski
Department of Electrical and
    Computer Engineering
University of Delaware
Newark, Delaware

Dennis W. Prather
Department of Electrical and
    Computer Engineering
University of Delaware
Newark, Delaware

Paul J. Resnick
MEMS Core Technologies
Sandia National Laboratories
Albuquerque, New Mexico

Garrett Schneider
Department of Electrical and
    Computer Engineering
University of Delaware
Newark, Delaware

Ahmed Sharkawy
Department of Electrical and
    Computer Engineering
University of Delaware
Newark, Delaware

and

EM Photonics, Inc.
Newark, Delaware

Shouyuan Shi
Department of Electrical and
    Computer Engineering
University of Delaware
Newark, Delaware

Jin Won Sung
Ostendo Technologies, Inc.
Carlsbad, CA

Jian (Jim) Wang
NanoOpto Corporation
Somerset, New Jersey

Joel R. Wendt
Photonic Microsystems
    Technologies
Sandia National Laboratories
Albuquerque, New Mexico

1 Fabricating Surface-Relief Diffractive Optical Elements

Shanalyn A. Kemme and
Alvaro A. Cruz-Cabrera

CONTENTS

Diffractive optical elements (DOEs) are exceptional optical devices that can deflect light into m orders at precise angles, θ_m. This is quantified in the diffraction equation:

$$n_{\mathrm{d}} \cdot \sin \theta_{dm} = \frac{m \cdot \lambda}{\Lambda} + n_i \cdot \sin \theta_i \qquad (1.1)$$

An incident ray, at θ_i with respect to the normal, upon a grating is diffracted into one or more departure angles, θ_{dm}, which is a function of the period, Λ. n_i is the

1

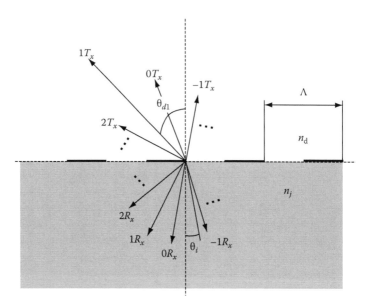

FIGURE 1.1 Depiction of the output angles as a function of grating period, refractive indexes, and order number. The diagram shows some of the transmitted and reflected orders.

refractive index of the incident medium, n_d is the refractive index of the departure medium, and m is the order of interest. For many applications, m is 1. Figure 1.1 indicates the existence of other orders, but not their associated efficiency. Often, for lenses, the goal is to minimize the efficiency of all but the desired orders, usually the $m = +1$ order.

The period is a parameter that is controlled laterally usually using lithographic methods, and can be specified down to the nanometer level. This exceptional control in the lateral position of the period makes DOEs ideal for low-aberration lenses. For example, a common application for a lens is focusing light to a spot; diffractive lenses can be designed and fabricated to obtain a diffraction-limited spot. Figure 1.2 depicts an 8-level focusing lens that stitches the phase front of an incident collimated beam every 2π to ensure that all the light diffracted by the DOE converges to a point centered above the DOE at a distance F. The phase front required to focus the collimated beam can be easily determined using an optical ray tracing tool. The location and the distance between each 2π discontinuity determines the period, Λ, at each position on the DOE. These values can be entered into a CAD program and transferred to a series of lithographic masks that will be used in the fabrication of the DOE. This process is complex but is well understood, and the technologies involved are mature.[1]

1.1 FABRICATION METHOD

The selected fabrication method and working within its limitations are the primary factors in obtaining desired DOE efficiency. For example, Figure 1.3 shows an ideal

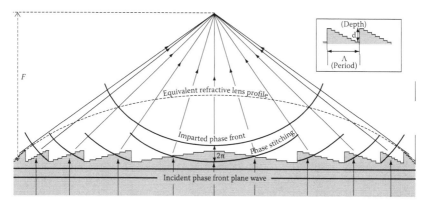

FIGURE 1.2   Depiction of an 8-level focusing lens that stitches the phase front of a collimated beam every 2π to ensure all light diffracted by the DOE is incident upon a point above the DOE at a distance F.

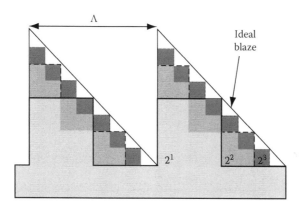

FIGURE 1.3   Depiction of 2-, 4-, and 8-level multilevel profiles inscribed in an ideal blazed grating. In the scalar regime, the ideal blazed grating is more efficient than any equivalent multilevel grating.

blazed grating of the period Λ, with possible multilevel alternatives to approximate the blazed profile, including 2-level, 4-level, and 8-level approximations. These multilevel profiles are based on a 2^N levels approach, where N is the number of masks.[2] A typical fabrication cycle has four steps (spin photoresist, expose and develop, etch, and clean) per mask. As the number of masks increases, the profile approaches the ideal blazed shape. For periods larger than 10λ, considered the scalar regime, the ideal blazed shape has the highest grating efficiency. In scalar theory, more levels (finer steps) correspond to increased efficiency as in Figure 1.4. However, the delta increase in efficiency diminishes as the number of masks increases and realized gains compete with fabrication errors. For example, the maximum delta increase in efficiency from 2-level to 4-level (one more mask) is 40.6%, whereas adding one more mask step from an 8-level to a 16-level realizes only 4%.

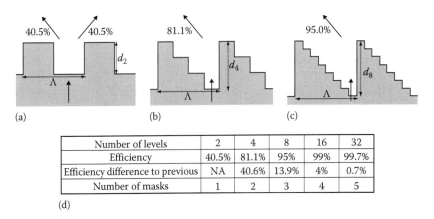

(a)   (b)   (c)

Number of levels	2	4	8	16	32
Efficiency	40.5%	81.1%	95%	99%	99.7%
Efficiency difference to previous	NA	40.6%	13.9%	4%	0.7%
Number of masks	1	2	3	4	5

(d)

FIGURE 1.4 Scalar regime efficiencies for a (a) 2-, (b) 4-, and (c) 8-level grating. Note that the depth, d_2, for the 2-level grating imparts a phase delay of π, while the total depth for any higher-level grating approaches a full 2π delay. The table (d) shows the scalar efficiencies for 2- to 32-level gratings and the efficiency delta with the added mask.

The ideal blazed grating can be fabricated with a single gray-scale mask and consequently does not have problems with misalignment between masks. The main drawback is that it requires a time-consuming development process due to the nonlinear responses of the photoresist, exposure, and the etch. Each of these process steps must be characterized exactly for the design parameters. If anything in the process should change, such as the design wavelength or the utilized photoresist, the entire process must be recalibrated. Because of the nonlinearity of these sequential steps, depth errors at any fabrication step can be significant. Nevertheless, fabricating the ideal blazed grating to specifications will guarantee the highest efficiency for a scalar period.

Even a 2-level phase grating ($N = 1$) is more efficient than a sinusoidal phase grating fabricated using the interference between two beams (Figure 1.5). With a

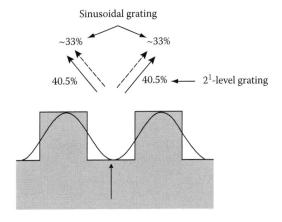

FIGURE 1.5 An example of a 2-level phase grating which is more efficient than a sinusoidal phase grating.

normally incident beam, a depth-optimized sinusoidal grating has a +1 and a −1 order with efficiencies of 33% each, while the efficiencies for the same two orders using a depth-optimized 2-level grating is approximately 40.5% each.

1.2 PERIOD AND WAVELENGTH RATIO

The first-order efficiency of a DOE can behave unexpectedly when the ratio between period and wavelength is between 1 and 10, especially when the ratio nears 2. This regime is known as the quasistatic or the vector regime where efficiencies should be calculated using the full vectorial form of Maxwell equations. Rigorous coupled wave analysis (RCWA) can be used to calculate the transmitted and reflected efficiencies for all propagating orders in a periodic grating. A first-order approximation for the total grating depth can be calculated using the following scalar equation:

$$d = \frac{(2^N - 1)\lambda}{2^N(n-1)} \tag{1.2}$$

This depth calculation is appropriate for gratings with periods larger than 10λ, the scalar regime. However, gratings with periods in the quasistatic regime and the optimum depth should be determined using RCWA. Following standard photolithographic techniques, all of the features within an etch cycle have the same design depth. Figure 1.6 shows efficiencies of first transmitted and second

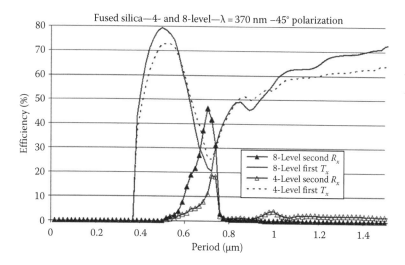

FIGURE 1.6 Efficiency curves calculated using RCWA, as a function of period, for the transmitted first and reflected second orders for 4- and 8-level gratings at a wavelength of 0.37 μm and at normal incidence. The graphs show a pronounced loss in efficiency at periods around 2λ (0.74 μm), where the 4-level transmitted efficiency is greater than the 8-level transmitted efficiency; and a corresponding increase in the reflected second-order efficiencies. Around a period of 1.35λ (0.5 μm), both first-order transmitted efficiencies are maximal, since any higher orders are evanescent and the geometry of the grating is optimized for the $m = +1$ transmitted order.

reflected orders for 4- and 8-level gratings as a function of the period. The gratings in Figure 1.4b and c are simulated in fused silica at a wavelength of 0.370 μm and a polarization of 45° to indicate that there is no preference for any polarization orientation. The fused silica refractive index is 1.474 at a 0.370 μm wavelength for a total scalar-predicted depth of 0.586 and 0.683 μm for the 4- and 8-level gratings, respectively. At larger periods, the first-order predicted efficiency approaches the scalar values. For periods less than λ (0.37 μm) the first order ceases to exist. From a period of 1.7λ (0.63 μm) to 2.4λ (0.9 μm), the efficiencies for both gratings drop below 50%. Even more unusual are the regions where the 4-level transmitted efficiency is higher compared to the 8-level transmitted efficiency. Around 2λ (0.74 μm), the transmitted efficiency for an 8-level grating drops to 20%, while for the 4-level grating it drops only to 25%. Correspondingly, the second-order reflected efficiency rises to 45% for the 8-level but only to 20% for the 4-level grating.

When the grating period is around 1.35λ (0.5 μm), both first-order transmitted efficiencies are maximal. At this period all higher orders are evanescent and the depth favors the +1 transmitted order over the −1 transmitted and ±1 reflected orders.

1.3 SHAPE OF THE GRATING

Maintaining the designed grating shape is critical to guarantee high transmission of the order(s) of interest. Etching to the optimal depth is essential since missed depths will appear as increased zero-order transmission. Other shape changes, such as those from mask misalignment, will scatter light to higher orders. A few shape deviations are not critical to transmitted efficiencies, like rounded profile edges, unless there is a pronounced asymmetry in the rounding. These effects are quantified in the following sections.

1.4 DEPTH OPTIMIZATION

As described in Section 1.3, gratings, and thus DOE lenses, have fixed depths across the device if they are fabricated using conventional methods. Generally, Equation 1.2 gives an optimal total depth, but with high numerical aperture lenses there are grating periods in the quasistatic regime, particularly near the edges of the lens. Figure 1.7 shows that at periods smaller than 3λ (4.65 μm), the scalar depths of 2.62 μm for 4-level and 3.05 μm for 8-level are not ideal. On lenses with a large distribution of periods smaller than 3λ (4.65 μm), these depths should be identified and specified using the full vectorial solution of Maxwell equations. The solid curve in Figure 1.8a shows the transmitted first-order efficiency calculated using RCWA at a fixed depth (from the scalar approximation) of 3.05 μm. This depth is also indicated in Figure 1.7b as a black line. The dashed curve in Figure 1.8a is the efficiency if the depth could be optimized for each period. Figure 1.8b shows the total depths corresponding to the efficiency curves in Figure 1.8a. The fixed etch depth is a solid line and the optimized etch depth is the dashed curve for each period. Note that for small periods, this depth can reach 7.5 μm. Finally, Figure 1.8c shows a possible example of the profile using these efficiency-optimizing depths.

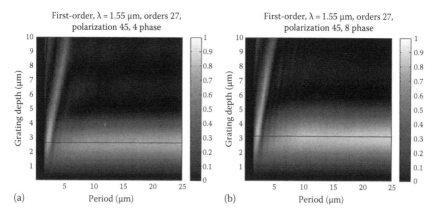

FIGURE 1.7 Transmitted first-order efficiency graphs for (a) 4- and (b) 8-level gratings in fused silica at a wavelength of 1.55 μm. The first-order efficiency is shown as a function of period (x-axis) and grating depth (y-axis). The efficiency is represented by a level of gray, where white is one and black is zero. The two graphs show that it is possible to set the overall depth of a DOE to a value close to the scalar depth (black lines at a depth of 2.62 μm for 4-level and 3.05 μm for 8-level) and achieve high efficiencies for many periods. However, in this case for periods smaller than 3λ (4.65 μm), the optimal depths are different from the scalar depths.

The greatest deviation from the scalar depth will occur when the period ranges from λ to 3λ (1.6–4.7 μm) as in Figure 1.8a where the minimum first order is approximately 55%, instead of 20%, for an 8-level grating at a wavelength of 1.55 μm. The depths to obtain these efficiencies could reach 6–7 μm as in Figure 1.8b and c. These depths may be impractical since these features are often located at the edges of a lens where there is minimal light to be diffracted, the periods are extremely small, and the impact of fabrication error is high.

A DOE with multiple periods, such as a lens, may exhibit depth nonuniformity as a result of the etch process. Figure 1.9 shows a scanning electron micrograph (SEM) image of a period-chirped grating in gallium arsenide (GaAs). All the periods were dry-etched for the same length of time, but they had smaller depths as the period decreased. The depth nonuniformity is a function of the type of the material being etched, the area density of the features, and the aspect ratio (depth/critical dimension) of the gratings. These gratings were realized using a chemically assisted ion-beam etcher (CAIBE). Fabrication of deep slots with subwavelength width and spacing presents a number of unique challenges. The need for slots of very high-aspect ratio with vertical walls dictates that a dry-etching process be used. The etch method must remove material from the bottom of the slot without removing material from the sides. Such anisotropic etching is accomplished with the CAIBE wherein the masked wafer is placed in a steady flow of reactant gas (Cl_2 and BCl_3 in our case) and exposed to a highly directional beam of energetic Ar ions. The Ar ions will physically knock GaAs atoms out of the wafer surface causing a sputter etch following the directional nature of the Ar ion beam. The addition of a reactant gas to the wafer surface reduces the binding energy of the Ga and As atoms at the surface allowing for improved etch rates and control of the sidewall angle.[3]

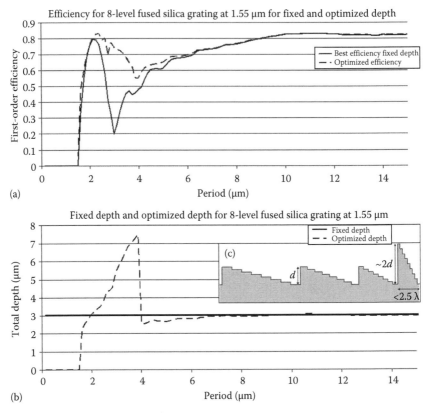

FIGURE 1.8 Panel (a) shows efficiency curves as a function of period for the transmitted first orders of an 8-level grating at a wavelength of 1.55 μm. The solid curve shows the efficiency for a fixed depth of 3.05 μm (from scalar approximation). The dashed curve is the efficiency if the depth is optimized for each period. Panel (b) shows the fixed etch depth (solid line) and the optimized etch depth (dashed curve) for each period. Note that for small periods, this depth can reach 7.5 μm. Inset panel (c) shows an example profile using the efficiency-optimizing total depth.

Figures 1.10 and 1.11 show examples of fabricated gratings in GaAs. The two devices were fabricated with different levels of control over the angular spread of the Ar ion beam. This was accomplished by adding an intermediate spatial filter to the etching beam. Figure 1.10 shows nonrectangular grating profiles that are a result of off-axis etch species. The spatially filtered etching beam limits a significant proportion of energetic Ar ions that are not normal to the wafer plane; blocking primarily the higher-angle sputtering species. This reduces the uniformity over a large wafer area but improves the rectangular grating profiles as in Figure 1.11.

The measurements, even with the improved spatial filtering, show a reduction in etch depth as the duty cycle increases. Figure 1.12 shows the slot depth (and slot bottom width) as a function of slot width at the top. The period ranges from 400 nm to 4 μm. The data with no added spatial filtering correspond to the large beam diameter

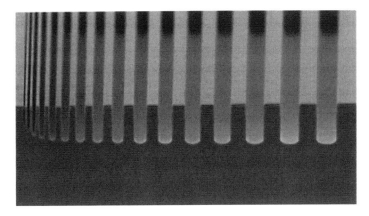

FIGURE 1.9 SEM of a CAIBE, chirped grating in GaAs. The resulting etch depths range from 1.25 to 2.65 μm as the period increases from 400 nm to 4 μm. (From Kemme, S.A. et al., *Proc. SPIE—Int. Soc. Opt. Eng.*, 6110, 112, 2006.)

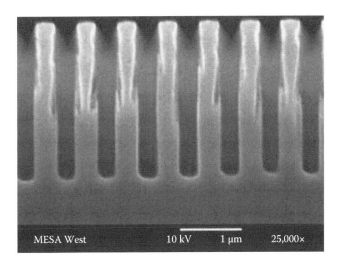

FIGURE 1.10 High-aspect ratio features in a grating in GaAs, affected by off-axis angle etch species.

squares, while the improved spatial filtering circles correspond to the small beam diameter data. Excluding the data point for a slot width of 100 nm (duty cycle of 90%), the grating depth varies by less than 20% over the duty cycle range.

1.5 MISALIGNMENT

Misalignment is one of the most significant issues, after depth, that affects the efficiency of a DOE. The problem is the overlay of a mask with respect to a previously placed mask position. Error in overlay can change the shape of a grating and in most

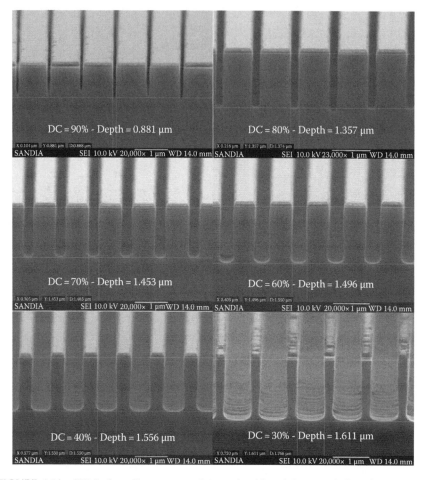

FIGURE 1.11 SEM of profiles across a duty cycle chirped 1 μm period grating made in GaAs. The images show the decreased depth for large duty cycles (DC) while preserving the period across the device.

cases will reduce the first-order efficiency. Figure 1.13a shows the fabrication steps to realize an ideally aligned 4-level grating whose efficiency can be easily predicted. Figure 1.13b depicts the same 4-level grating process where the second mask is misaligned toward the negative axis (left shift). This error can create thin and deep trenches that may be transferred into the substrate if the etch is highly anisotropic. Similarly, Figure 1.13c shows a shift toward the positive axis (right shift) that creates tall and thin features, which may be transferred into the etch but are quite fragile. Figure 1.14a shows the SEM of an 8-level grating where two thin and tall (~0.1 μm wide and 0.3 and 1 μm tall) features survive fabrication and cleaving. Figure 1.14b shows the idealized representation of the misalignment in Figure 1.14a with the second and third masks misaligned toward the positive axis (right shift). Notice that the third feature did not survive at this cleaved section but may exist in other locations on the grating.

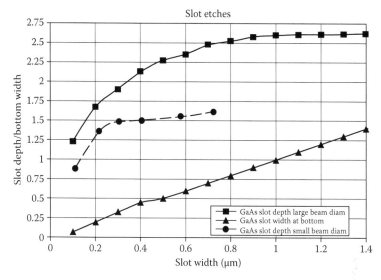

FIGURE 1.12 Measured etch depths and slot widths for chirped gratings in GaAs for two different etching beam diameters. The period ranges from 400 nm to 4 μm. The data with no added spatial filtering correspond to the large beam diameter squares while the improved spatial filtering circles correspond to the small beam diameter data.

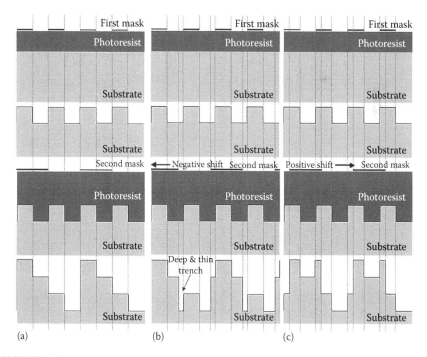

FIGURE 1.13 (a) Basic process for fabricating a properly aligned 4-level grating. (b) Misalignment of the second mask toward the negative axis (left shift) resulting in deep, thin trenches and high shoulders. (c) Misalignment of the mask toward the positive axis (right shift) resulting in tall, thin features and low shoulders.

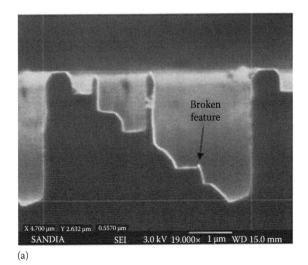

(a)

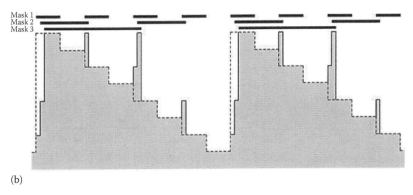

(b)

FIGURE 1.14 (a) Misalignment in an 8-level grating of the second and the third mask toward the positive axis with respect to the first mask. The second tall structure is an artifact of the third mask misalignment in the same direction as the second mask (positive axis). (b) Diagram of an equivalent 8-level grating and the shifted masks.

The decrease in first-order efficiency due to the positive shift of the second mask in a 4-level fused silica grating is modeled in Figure 1.15a. The grating in Figure 1.15b has a period of 5 μm (or ~3.22λ) for a wavelength of 1.55 μm and a refractive index of 1.444. This grating is a part of a radially symmetric lens and a canonical 10λ (15.5 μm) period determines the depth of the entire lens. The total depth of the grating is 2.59 or 0.863 μm per step, and is simulated using RCWA. The linear polarization of the incident light is at 45°, modeling no preference for TE or TM. The analysis shifts the second mask up to 20% of the period or 1 μm. The grating has 7 transmitted and 9 reflected orders. With no shift, the +1 transmitted order efficiency is 61% and the −1 transmitted order efficiency is 9%. With a mask shift of 0.2Λ, the +1 transmitted order drops to 10% and the −1 transmitted order increases to 50%. The switch in efficiencies occurs with a mask shift around 0.12Λ. With the other 5 transmitted orders, only 0, +3, and −3 reach 12% across different mask shifts. The reflected order efficiencies are negligible.

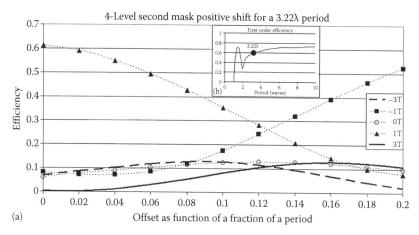

(a)

FIGURE 1.15 (a) +1 transmitted order efficiency as a function of the second mask positive direction misalignment with respect to the first mask for a 4-level grating. The grating has a period of 5 μm and a wavelength of 1.55 μm (3.22λ as in (b)) indicating several more transmitted orders. As the shift increases, the light is transferred to the −1 order reaching 50% at a shift of 20% of the period. The next most significant change in efficiency is for the −3, 0, and 3 transmitted orders, each one reaching 12% at different offsets.

Figure 1.15b shows the +1 transmitted efficiency as a function of the period of the grating. The depth of the grating determines the efficiency. For many applications, other gratings in the DOE are in the scalar regime and they dictate the total depth of the device. These gratings have more than 3 transmitted orders to transfer light when there is a misalignment. However, in some cases the DOE has gratings only with periods smaller than 2λ. In this case, the optimal etch depth of the gratings can exceed their scalar depth (see Figure 1.8b), and there are only the +1, 0, and −1 transmitted orders (and a similar number of reflected orders) to transfer light.

A similar analysis is done for the first-order efficiency but with a negative shift of the second mask for the same 4-level fused silica grating, and is shown in Figure 1.16. The analysis shifts the second mask up to −0.2Λ. With the full 0.2Λ mask shift, the +1 transmitted order drops to 15% and the −1 transmitted order goes up to 48%. With the other 5 transmitted orders, only +3 and −3 transmitted orders reach 10% across different mask shifts and the 0 transmitted order is maintained between 5% and 6%. The reflected order efficiencies are still negligible.

Figure 1.17a shows a similar analysis but for a different grating. Here, the grating has a period of 1.3 μm, or ~1.5λ (see Figure 1.17b), for a wavelength of 0.86 μm and a refractive index of 1.452. The total depth of the grating is 1.66 or 0.553 μm per step, and is simulated with RCWA for a period of 1.5λ. The scalar depth of the grating is 1.43 μm. The linear polarization of the incident light is at 45° without any preference for TE or TM. The analysis shifts the second mask in the positive direction up to 20% of the period (0.2Λ) or 0.26 μm. The grating has 3 transmitted orders and 5 reflected orders. With no shift, the +1 transmitted order efficiency is 80% and the −1 transmitted order efficiency is less than 5%. With a mask shift of 0.2Λ, the +1 transmitted order drops to 11% and the −1 transmitted order goes

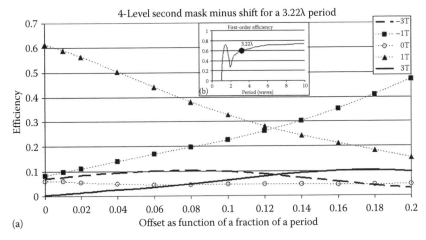

FIGURE 1.16 (a) +1 transmitted order efficiency as a function of the second mask negative direction misalignment with respect to the first mask in a 4-level grating. The grating has a period of 5 μm and a wavelength of 1.55 μm (3.22λ as in panel (b)) indicating several more transmitted orders. As the shift increases, the light is transferred to the −1 order reaching 48% at a mask shift of 20% of the period. The next most significant change in efficiency is for the −3 and 3 transmitted orders, each one reaching 10% at different offsets. The zeroth transmitted order efficiency is between 5% and 6% across different offsets.

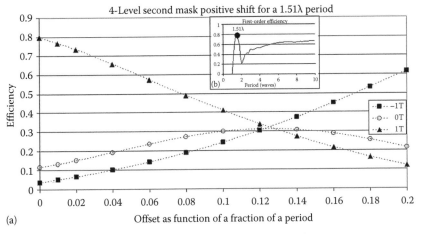

FIGURE 1.17 (a) +1 transmitted order efficiency as a function of the second mask positive direction misalignment with respect to the first mask for a 4-level grating. The grating has a period of 1.3 μm and a wavelength of 0.86 μm (1.5λ as in panel (b)) indicating a lack of higher transmitted orders. As the shift increases, the light is transferred to the −1 transmitted order reaching 60% at a shift of 20% of the period. The light transferred to the zero transmitted order reaches 31% at 12% of the period shift.

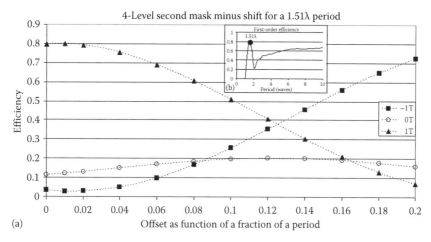

FIGURE 1.18 (a) +1 transmitted order efficiency as a function of the second mask negative direction misalignment with respect to the first mask for a 4-level grating. The grating has a period of 1.3 μm and a wavelength of 0.86 μm (1.5λ as in panel (b)) indicating a lack of higher transmitted orders. As the shift increases, the light is transferred to the −1 transmitted order reaching 70% at a shift of 20% of the period. The light transferred to the zero transmitted order only reaches 20% at 10% of the period shift.

up to 61%. The zero transmitted order reaches 30% around 0.12Λ. The reflected order efficiencies are negligible.

Finally, Figure 1.18 shows the equivalent analysis for the same grating of the first-order efficiency but with a negative shift of the second mask. The analysis shifts the second mask up to −0.2Λ. With the full 0.2Λ mask shift, the +1 transmitted order drops from 80% to 8% and the −1 transmitted order goes up from more than 5% to 72%. The zero transmitted order reaches 20% around 0.12Λ. The reflected order efficiencies are negligible.

In both the long-wavelength and shorter-wavelength cases described, we see the onset of a pronounced drop in the efficiency of the desired order after a misalignment of only 2% of the grating period. We, therefore, must reduce the misalignment error to be within this margin. There are multiple approaches to reduce this error.

Misalignment error is controlled by placing fiducials on the peripheries of the mask and using them to minimize the shift in the x- and y-axes, and the rotation. A minimum of two fiducial sets is needed for alignment and they should be placed at opposite edges of the mask. It is prudent to place other sets about the mask to mitigate the poor patterning yield in the fiducials.

Two approaches can be used in the alignment of the masks. The first method references each mask to the first etched level. The main disadvantage to this approach is that the first etched features may not be well defined, and if the etch is shallow, the contrast is too small to allow for proper alignment through the photoresist when patterning the next mask. The contrast issue is problematic with

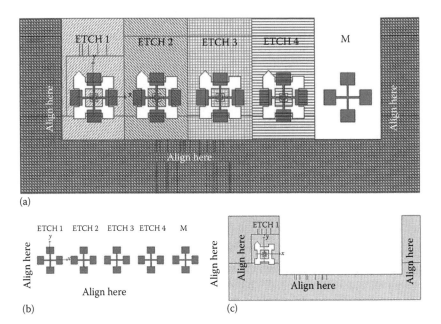

FIGURE 1.19 (a) Full alignment fiducial set for a 16-level grating and a metal reference layer. (b) Metal reference layer, usually gold (1000–2000 Å thick) for good contrast. (c) The first-etch fiducial overlay; the gray area is open and will be etched in a positive photoresist process.

fused silica because the refractive index difference between fused silica and the photoresist is small.

The second technique uses a metal reference layer. The reference layer patterns only the fiducials through liftoff or leaves the wafer open for the DOE. The primary benefit of this technique is that it is possible to obtain a good contrast between the substrate and the fiducials, which can be maintained during all the fabrication steps. Figure 1.19a depicts a fiducial mask for a 16-level grating plus an additional metal layer. The reference layer, shown in Figure 1.19b, is a metal cross pattern that should have good contrast with respect to the substrate. Figure 1.19c shows the first-etch alignment fiducial that overlays the metal reference cross. The gray area is open and is etched, destroying the reference cross. The fiducial is patterned to leave a protective layer of the photoresist on top of the reference crosses used in the alignment of the subsequent masks. The thin lines in the reference crosses are 2 µm thick, allowing for less than 1 µm misalignment using contact lithography. An e-beam writer can minimize overlay errors to less than 0.1 µm.

1.6   EDGE ROUNDING

There are two process steps that can contribute to rounded grating edges: the masking step and the etching step. The first cause of rounding is the thickness nonuniformity of the applied mask, particularly with photoresist masks. Secondly, etching may

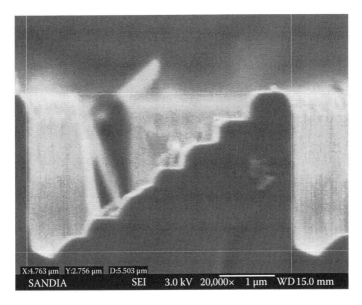

X:4.763 µm   Y:2.756 µm   D:5.503 µm

SANDIA   SEI   3.0 kV   20,000×   1 µm   WD 15.0 mm

FIGURE 1.20   SEM of 8-level, 4.8 µm period grating with rounding at each step.

not be as anisotropic as desired, leading to the rounding of sharp corners. Figure 1.20 shows that the steps, originally etched with the first mask, become more rounded and smooth with subsequent etches. These steps were exposed repeatedly to the reactive ion etch (RIE) plasma. The rounding is pronounced for large aspect ratios where the RIE must remove material from deep inside trenches. The effect on optical performance of these fabrication issues is quantified below.

For this analysis, we use RCWA and approximate the rounded corners with a circular profile of radius R_x with stacked rectangles. The rectangles have a variable thickness and a constant delta in the lateral dimension of $0.1R_x$ (see Figure 1.21a). A radius is selected and optical efficiency is simulated. Figure 1.21b shows the grating profile for a range of radii. The upper limit for the radius is the height or the lateral dimension of the feature, whichever is smallest. Half of the lateral dimension is considered if the rounding occurs at both sides of the feature. The delta in the lateral dimension is at least 10 times smaller than the wavelength.

The analysis considers the efficiency of the +1, 0, and −1 transmitted orders as a function of the ratio between the total modified area and the total area of the original grating. The modified area is any area added to or eliminated from the original grating profile. Figure 1.22 indicates the modified area with diagonal black lines in white background and the original area under the solid and thick black line.

The first analysis is for 2-level fused silica gratings. The original grating has a depth of 1.94 µm, a period of 2.55 µm, with a duty cycle of 50%. The wavelength is 1.55 µm and is at 45° polarization, and the transmitted efficiency for the +1 or the −1 order is 43%, and the transmitted zero-order efficiency is 9.7%. We analyze symmetrical rounding of the top in Figure 1.23a, rounding at the top and bottom in Figure 1.23b, ditches at the bottom in Figure 1.23c, and rounding at the top and ditches at the

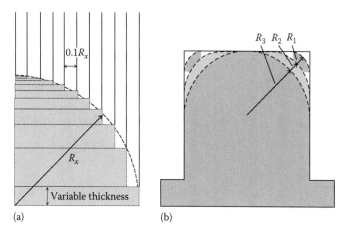

(a)   (b)

FIGURE 1.21   (a) Layered rectangles used to model the profile corner rounding using RCWA. The rounded corners are approximated using rectangles with variable thickness and fixed delta in the lateral dimension. This delta is defined as 1/10 of the radius. (b) The analysis looks at the efficiency across a range of radii with the height and/or the lateral dimension of the feature as an upper limit.

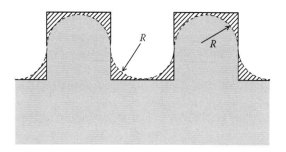

FIGURE 1.22   Area modified by the rounding is the area with diagonal black lines in white background, and the original area under the solid and thick black line.

bottom in Figure 1.23d. The efficiency barely changes for all the cases with the worst case being the rounding at both the top and bottom, as in Figure 1.23b, where the +1 or −1 order efficiencies drop to 40.7% and the zeroth-order efficiency increases to 12.7%. This case is equivalent to etching the grating to 1.67 μm instead of 1.94 μm. For the other three cases a similar behavior is seen but to a lesser degree.

The next analysis uses the same original grating but the modification to the profile is asymmetric. One of the devices has rounding at both sides of the top, but one side has half the radius compared to the other side as in Figure 1.24a. In the second case, rounding at the top occurs only at one side as in Figure 1.24b. In the third case, there is a ditch on only one side (see Figure 1.24c). The last case is unusual but may be related to the quality of the glass and it appears as pitting along one edge of the grating profile, as in Figure 1.24d.

The third analysis involves 4-level gratings in fused silica with a 4 μm period and a total depth of 2.5 or 0.833 μm per step. The incident light has a wavelength of 1.55 μm and is at 45° polarization. This grating has the lateral dimension for the first step optimized to 0.2Λ or 0.8 μm. The other steps' lateral dimensions are 0.3Λ, 0.25Λ, and 0.25Λ. This change improves the efficiency from 54% to 57% with minimum effort in the fabrication. The analysis looks at rounding of the first and third steps, both in the same direction, as in Figure 1.25a. The radius of the curves at each step reaches 0.8 μm or 0.2Λ of the period. For the next two cases, the rounding is symmetric at both sides of the first step and the radius reaches only 0.4 μm for all steps.

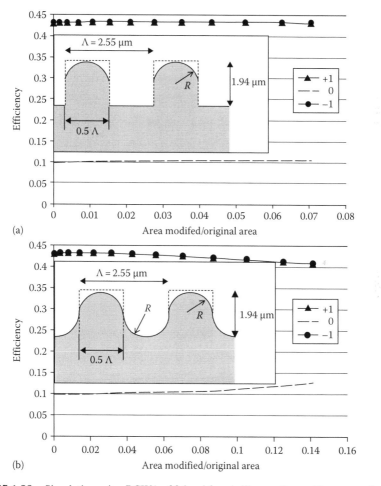

(a)

(b)

FIGURE 1.23 Simulation using RCWA of 2-level fused silica gratings with symmetric profile deviations. The incident light has a wavelength of 1.55 μm and is at 45° linear polarization. The graphs shows +1, 0, or −1 transmitted order efficiency as a function of the ratio between the area modified and the original area of the grating. Gratings analyzed: (a) rounding at the top, (b) rounding at the top and bottom, (c) ditch with rounding at the bottom, and (d) ditch with rounding at the bottom and rounding at the top.

(continued)

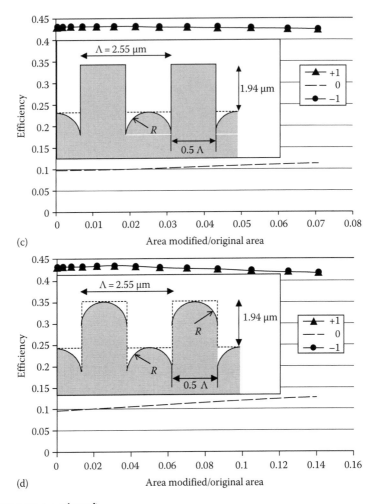

FIGURE 1.23 (continued)

One analysis rounds the third step only, as in Figure 1.25b, and the next one rounds the second and third steps, as in Figure 1.25c. The results indicate small changes in efficiency, where the largest change seen is in the first case (Figure 1.25a).

The last analysis involves 8-level gratings in fused silica with an 8 μm period and a total depth of 2.93 or 0.4186 μm per step. The incident light has a wavelength of 1.55 μm and is at 45° polarization. The analysis looks at rounding of the first, third, fifth, and seventh steps, all in the same direction as in Figure 1.26a. The radius of the curves at each step terminates at 0.4186 μm or at the depth of the steps. For the next case, the rounding is symmetric at both sides of the first step and the radius is limited to 0.1864 μm, the same as in the previous case. This configuration rounds the third, fifth, and seventh steps as in Figure 1.26b. The results indicate an even smaller change in efficiency than in previous cases.

The general finding from these rounding analyses is the insignificant drop in efficiency, indicating that roundings do not fundamentally change the overall blazed geometry for the 4- and 8-level gratings. Etch depth and misalignment errors have a more detrimental effect on efficiency than rounding, and a higher effort should be invested in controlling those parameters than in controlling rounding. In the 2-level grating, there are some changes in the efficiency with symmetric profile modification. Only when the symmetry of the 2-level grating shape is distorted do we see significant changes in efficiency; the efficiency is

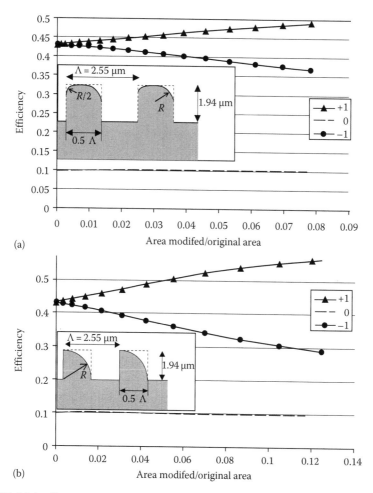

(a)

(b)

FIGURE 1.24 Simulation using RCWA of 2-level fused silica gratings with asymmetric profile deviations. The incident light has a wavelength of 1.55 μm and is at 45° linear polarization. The graph shows +1, 0, and −1 transmitted order efficiency as a function of the ratio between the area modified and the original area of the grating. Gratings analyzed: (a) uneven rounding at the top, (b) one side rounding at the top, (c) ditch with rounding at one side, which can be seen in the SEM, and (d) along one side of the 50 nm thick grating profile.

(*continued*)

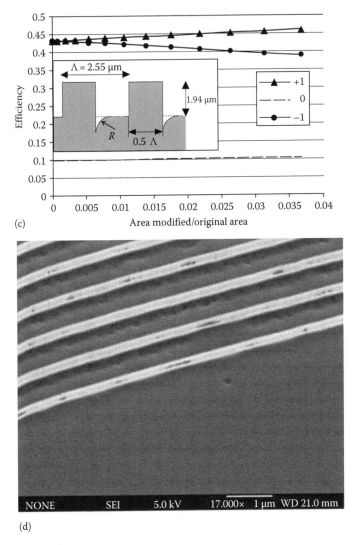

(c)

(d)

FIGURE 1.24 (continued)

redistributed between the +1 and −1 transmitted orders, which are between 10% and 20%, with a change of 3.5% in shape.

1.7 CHANGES IN PHASE RESPONSE OF FORM-BIREFRINGENT GRATINGS DUE TO GEOMETRIC DEVIATIONS

An analysis is done for the induced phase delay of a form-birefringent grating with rounding in the trenches. Gratings in GaAs were fabricated with a target period of 0.65 μm, a duty cycle of 80% for the lines (0.52 μm), and a depth of 1.23 μm.

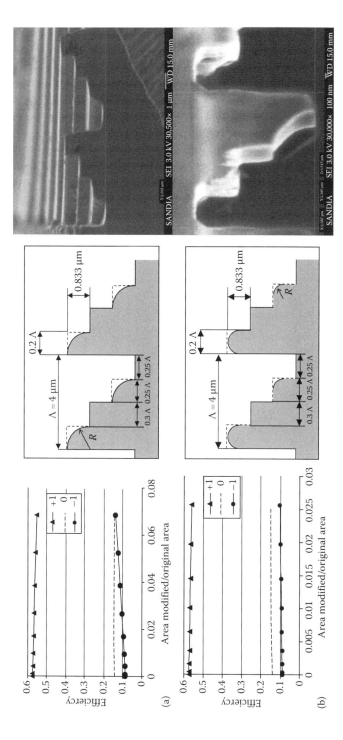

FIGURE 1.25 Simulation using RCWA of 4-level fused silica gratings with profile deviations. The incident light has a wavelength of 1.55 μm and 45° linear polarization. The graph shows +1, 0, and −1 transmitted order efficiency as a function of the ratio between the area modified and the original area of the grating. Gratings analyzed: (a) one side rounding on the first and third steps, (b) even side rounding on the first and third step, (c) even side rounding on the top, and the second and third step.

(continued)

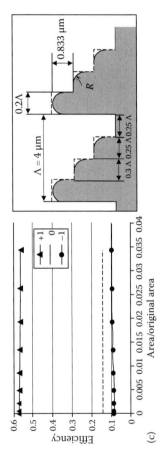

FIGURE 1.25 (continued)

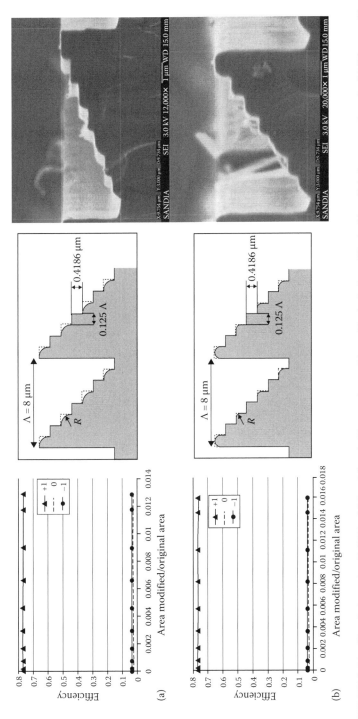

FIGURE 1.26 Simulation using RCWA of 8-level fused silica gratings with profile deviations. The incident light has a wavelength of 1.55 μm and is at 45° linear polarization. The graphs shows +1, 0, and −1 transmitted order efficiency as a function of the ratio between the area modified and the original area of the grating. Gratings analyzed: (a) one side rounding on the first, third, fifth and seventh step, (b) even side rounding on all steps.

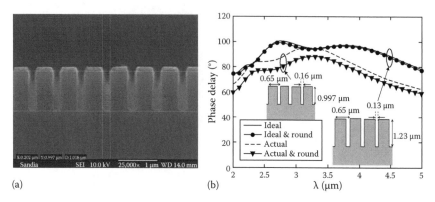

(a)   (b)

FIGURE 1.27 (a) SEM of form-birefringent grating in GaAs. The grating shape diverges from target depth, duty cycle, and square profile at the bottom. (From Boye, R.R. et al., *J. Microlith. Microfab. Microsyst.*, 5(4), October–December, 2006.) (b) Phase delay of the grating for a linearly polarized incident light at 45° with respect to the grating grooves, across a broad wavelength span of 3 µm. The graph contains the phase delay response for the targeted design, the effect of rounding on the target design, an actual device based on measurements from the SEM, and an actual device incorporating rounded trenches in the grating.

The form-birefringent gratings are designed for the wavelength range of 2–5 µm. The as-fabricated device had a slight change in geometry, as shown in Figure 1.27a, where the actual duty cycle was 76% and the actual depth was 0.997 µm. The predicted phase delays are shown in Figure 1.27b as the solid line (target) and dashed lines (actual) without markers. Rounding in the grating trenches, with a radius of 0.065 µm for the target and 0.08 µm for the actual device, gives a slight reduction in the phase delay in both cases. The rounding effects decrease the induced phase delays between 0.3°–1.7° from the targeted design and 3.3°–7.8° from the actual, as-fabricated response. The primary changes in induced phase delays are imparted by the variations in the depth and the duty cycle.

1.8 SURFACE TEXTURING

The fabrication and the optical response of several DOEs highlight the relative importance of the subwavelength surface texture in the components' performance.[6] This surface texture is in addition to the larger, anisotropic DOE features that manipulate the propagating orders, and is commonly referred to as grass. The surface texture of an amorphous or a multicrystalline material is readily apparent in a SEM and is often an unavoidable consequence of the reactive ion etch (RIE) process. The contributing factors are mask erosion, self-masking, and material nonuniformity. We describe and quantify the effects of unavoidable and deliberate subwavelength surface texture on optical components.

One case quantifies the effect of the illustrated surface textures and indicates that some inadvertent subwavelength surface structures can in fact be beneficial. The next case describes solar cells where subwavelength texturing is intentional. Multicrystalline silicon wafers were utilized in an effort to lower solar cell material costs. However, because the material was no longer a single crystal, traditional wet-etch texturing processes that reduce the light lost to reflection were unavailable. Instead, large area, subwavelength, periodic and random RIE surface patterns were investigated. In this way, we implemented an effective antireflection (AR) coating with a response across a wide wavelength band using the subwavelength surface texture.

1.9 FUSED SILICA SURFACE STRUCTURES

Fused silica, in many ways, is an ideal material for realizing DOEs; it is highly transparent across a large wave band, it has a low refractive index that makes it optimal for cladding layers, and it has a low Fresnel reflection loss due to this low refractive index. At the same time, fused silica's low index contrast necessitates high-aspect ratio gratings that can be problematic for this low etch-rate, amorphous material. When the required absolute etch depth is small a well-behaved mask, such as poly(methyl methacrylate) (PMMA) that does not promote self-masking effects, gives rise to the desired DOE profiles. This is shown in the 4-level optical interconnect lens SEM picture on the left side of Figure 1.28. These good results are obtained with uniform, high quality fused silica.[7] The fused silica surfaces go through a critical solvent clean, hexamethyldisilazane (HMDS) spin, and are baked to remove the residual polishing compound and water before the mask application. We do not reflow the mask prior to etch, as is often done, instead relying upon a precise lithographic mask definition of smooth mask sidewalls to be transferred into the fused silica. The RIE etch process is fairly standard with 40 sccm of CHF_3, 3 sccm of O_2 at a pressure of 40 mTorr, a bias of 396 V, and 200 W RF power.

The inhomogeneous substrate material and/or self-masking conditions may produce an unintentional surface texture. The right-side SEM in Figure 1.28 illustrates the resulting etch into a lower quality fused silica substrate with an electron-beam-assisted deposition of SiO/SiO_2 material. The material is full of voids from the outset and a rough surface texture, grass, is the result.

FIGURE 1.28 SEM of optical interconnect lens with smooth surfaces etched in a high-quality fused silica substrate (left), and SEM of fused silica surfaces etched in a lower quality substrate with deposited SiO/SiO_2 material (right). (From Kemme, S.A. et al., *Proc. SPIE—Int. Soc. Opt. Eng.*, 5347, 247, 2004.)

While the appearance of grass is undesirable in a SEM, its effect upon optical parameters in a light management (nonimaging) optical system is not only minimal, but may in fact be advantageous. Here, we investigate the effect of a binary grating etched into fused silica, as illustrated in Figure 1.29. Such a grating may be modeled as an effective index layer, assuming that the period of the grating is well below the wavelength of light inside the fused silica. RCWA models the orders reflected and transmitted from the grating, and is used here to model the effect of the grating over a wide range of grating periods, from subwavelength to much larger than the wavelength.

Treating the grating layer as a uniform layer of material with a refractive index between that of air and fused silica is only valid in the regime where the grating period is less than the wavelength of light in the higher index material, in this case, fused silica. Increasing the depth of this layer (moving up in a vertical path in the two-dimensional left plot of Figure 1.30) shows a sinusoidal variation, as expected. The thickness of the layer that achieves the lowest reflection depends also on the refractive index of the layer. The left plot in Figure 1.30 shows the reflectivity from a fused silica surface that has been coated with a thin film of indicated refractive index. Regions of almost zero reflectivity can be seen for certain values of the refractive index and the depth of the thin film layer resulting in an AR coating.

Replacing the uniform thin-film layer with a subwavelength period, binary grating ($\Lambda = \lambda/2$) results in a near-identical plot (see the right plot in Figure 1.30) when the duty cycle is scanned instead of the refractive index of a thin film. These figures illustrate the equivalence between a thin film layer and a subwavelength grating. The effective index of the grating may be calculated for a specified incident light polarization, using the refractive indices of the two materials and the grating duty cycle. Thus, it is straightforward to precisely specify the reflectivity of an effective AR coating by choosing the appropriate duty cycle and depth of the binary grating.[8,9] It is clear that the surface texture, so long as the feature sizes remain smaller than the incident wavelength of light, will never increase the light lost due to reflection, and may make more light available for transmission.

However, there is no one-to-one correspondence between a homogeneous thin film and a subwavelength effective-index grating, as seen in Figure 1.31. As the period of the grating from the subwavelength regime increases, additional transmitted and reflected orders appear. Figure 1.31 may be broken down into three regions. The first is the region where the period of the grating is subwavelength compared to

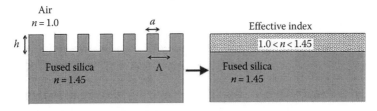

FIGURE 1.29 Subwavelength ($\Lambda < \lambda$) binary grating in fused silica and effective index equivalent. (From Kemme, S.A. et al., *Proc. SPIE—Int. Soc. Opt. Eng.*, 5347, 247, 2004.)

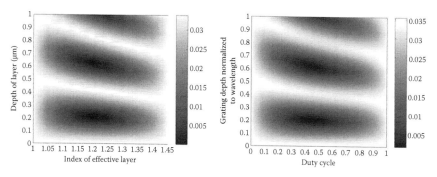

FIGURE 1.30 Reflectivity from a fused silica surface with a thin film (left) or a subwavelength ($\Lambda = \lambda/2$) grating (right) as a function of effective index or grating duty cycle for a $45°$ polarization, normally incident plane wave. (From Kemme, S.A. et al., *Proc. SPIE—Int. Soc. Opt. Eng.*, 5347, 247, 2004.)

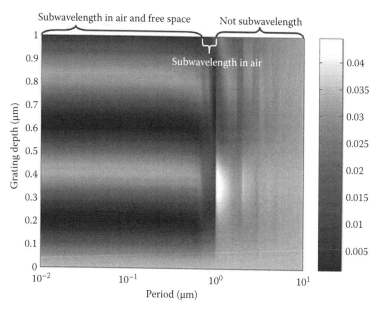

FIGURE 1.31 Reflectivity from a fused silica surface as a function of the grating period and grating depth for a $45°$ polarization, normally incident plane wave.

the wavelength of light in both air and fused silica, roughly for periods less than 0.7. In this subwavelength regime, we see no reflectivity variation with period and we see a sinusoidal reflectivity variation as a function of the grating depth. In the region, where the period varies from 0.7 to 1.0, the grating is a subwavelength compared to the wavelength in air, but is not a subwavelength in fused silica, thus allowing higher transmitted orders with only the zeroth-order reflected wave. In this regime, the reflectance is quite small and less dependent upon grating depth. Finally, in the range where the period is greater than the free space wavelength, higher reflected and transmitted orders may propagate.

1.10   SOLAR CELL TEXTURING

The goal of another surface texturing program was twofold: to reduce light lost in reflection, and to increase the red response by coupling the oblique incident light into transmitted propagating orders near the solar cell junction so that this light is absorbed before it exits the silicon wafer. A treatment of the solar cell surface is necessary in order to decrease the substantial light lost to the Fresnel reflection, as high as 32%, leading to a corresponding reduction in photocurrent. Conventional single crystal silicon may be cost-effectively textured with a wet-chemical process that relies on the uniform crystalline orientation.

Recently, the quality of the multicrystalline silicon has improved to the point that solar performance is close to that of single crystal silicon. However, this low-cost substrate cannot benefit from the inexpensive wet-etch surface treatment since the polycrystalline grains are randomly oriented. A maskless reactive ion etching (RIE) technique was developed[10] to obtain a wide spectral bandwidth, low reflection performance that is suitable for mass production. SEMs of random texture obtained from the maskless RIE process on a multicrystalline silicon are shown in Figure 1.32. The maskless RIE process for a multicrystalline silicon is better at reducing reflected light than the traditional wet-textured process for a single crystal silicon. This process involves several metal-catalyst-assisted RIE-texturing techniques using the SF_6/O_2 plasma chemistry. The cathode in the plasma chamber is constructed from graphite while the chamber walls and anode are made of aluminum. A large parameter space of power, pressure, gas ratio, flow rate, and etch time was investigated. A narrow parameter range was found to be useful for texturing single-crystal

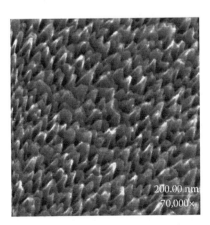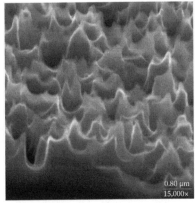

FIGURE 1.32   SEM pictures of the random texture obtained from the maskless RIE process on multicrystalline silicon surfaces. The scale bar on the left picture is 200 nm and that on the right picture is 800 nm. (From Kemme, S.A. et al., *Proc. SPIE—Int. Soc. Opt. Eng.*, 5347, 247, 2004.)

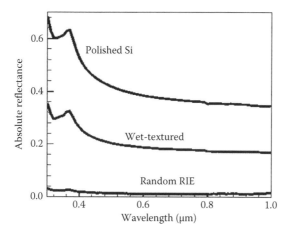

FIGURE 1.33 Hemispherical reflection spectrum measurements for a polished silicon surface, a wet-textured single crystal silicon surface, and a maskless RIE-textured multicrystalline silicon surface. (From Kemme, S.A. et al., *Proc. SPIE—Int. Soc. Opt. Eng.*, 5347, 247, 2004.)

silicon wafers up to 6 in. in diameter and multicrystalline silicon wafers of 130 cm² in area. The measured reflection spectra for two texturing processes, compared to a polished silicon surface, are shown in Figure 1.33.

Texturing produces a surface-relief structure with feature sizes smaller than the incident wavelengths of interest (400–1000 nm). This subwavelength structure acts as an effective AR coating. A thin-film AR stack is an alternative surface treatment for solar cells, but the broad spectral response and the angular tolerance of the single material silicon nanostructured surface is generally superior.

The overall optical response of the RIE nanostructuring is due to several parameters: etch depth, spatial frequency content, and subwavelength feature profile or shape. Our approach in this program was to model and analyze this RIE nanostructured surface as a linear combination of single-spatial frequency gratings and to investigate the effect of each of these parameters upon the total reflection spectra.

Experimental spectral measurements and numerical modeling of surface-relief single-spatial frequency silicon gratings were performed. Numerical modeling utilized RCWA to determine if the experimental results of one-dimensional gratings in silicon could be verified and possibly predicted. Reflected light was experimentally measured using a modified hemispherical reflectance system so that the polarization of the incident light was known. Incident TE light has the electric field polarization vector along the grooves of the grating. The vector for TM polarized light is perpendicular to the grating grooves. Across the surface of the wafer, there were slight variations in the duty cycle and the depth of the gratings. Since the measured data were taken over a relatively large area, the data are an average of grating depths and duty cycles found over this region. As a result, a single set of grating parameters was not sufficient to describe the experimental data, and so, an averaging method was employed.

From the SEM photographs of the gratings, a rough estimate of the grating depth and the duty cycle may be deduced. These estimates were used as starting points for determining the values for the modeling parameters that best fit the measured data. Modeling only a single estimate for the depth and the duty cycle results in plots of reflectivity versus wavelength with much finer features than the experimental data. This is to be expected as the measured data is averaged over an area with varying grating conditions.

An average of the calculated data was taken around the central point in order to smooth the curves. A ±5% range was chosen for the depth and the duty cycle such that the feature shapes of the averaged calculated reflectivity, as a function of the wavelength, closely resembled those of the experimental plot. In general, wider ranges lead to greater smoothing of the calculated curves when the four points are averaged.

Figure 1.34 shows a SEM of the fabricated triangular grating, dry-etched into silicon, and Figure 1.35 contains the modeled and measured results. This triangular grating demonstrates the lowest total reflectivity across the large wavelength band of the one-dimensional, single-spatial frequency fabricated gratings.

With this analysis tool, we can isolate and prioritize grating parameters that are primary contributors to the AR response, such as the profile shape, the period, and the depth. However, it is clear that a range of spatial frequencies is the key to achieving minimal reflection across a wide spectral band, as demonstrated by the randomly-etched texture measured in our maskless RIE process.

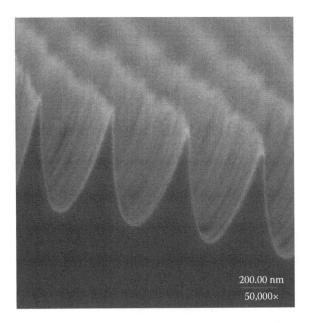

FIGURE 1.34   SEM picture of 420 nm period silicon triangular grating. (From Kemme, S.A. et al., *Proc. SPIE—Int. Soc. Opt. Eng.*, 5347, 247, 2004.)

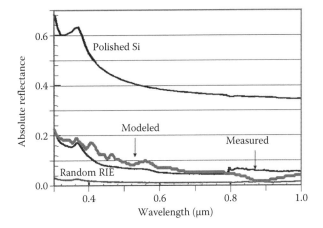

FIGURE 1.35 Measured (left) and calculated (right) spectra for the small 420 nm period triangular grating for randomly polarized light.

1.11 SAMPLE FABRICATION RECIPE FOR AN 8-LEVEL DOE IN FUSED SILICA

This is an example of a recipe for the fabrication of an 8-level fused silica lens. The first step is to define the desired etch levels for the design. Depth determination is dependent on the grating period distribution, intensity profile illuminating the lens, material constraints, and fabrication capabilities. In this case, we have a lens at a wavelength of 1.55 μm, where the refractive index is 1.44402. The DOE is a radially symmetric lens with 60% of the area falling within the scalar regime. We select the etch depth using the scalar equation (Equation 1.2) for a total depth of 3.05 μm, with a first depth of 0.43 μm, a second of 0.87 μm, and the third one of 1.75 μm. RCWA can be used to calculate optimal depths for the whole lens, but here, the differences in etch depths are in the order of 50 nm, which is near the typical fabrication error. The lens has 20% of the area working in the vector regime where periods range from 1.5λ to 5λ or 2.325 to 8 μm. The lens portion containing periods between 1.5λ and 3λ is fabricated as 4-level instead of 8-level. This eases the fabrication and still maintains good efficiency. We utilize the e-beam writer to reduce the misalignment errors to less than 0.1 μm since the lens has many features that are smaller than 1 μm.

A high quality glass should be selected for the fabrication. The glass should meet the scratch/dig and flatness requirements for the contact aligner. The e-beam writer has requirements for flatness but they are less stringent than those for the contact aligner. The main concern with the contact aligner is the ability to pull vacuum with the mask and a bowed substrate will slow if not impede good.

Two contact masks should be designed and procured: The metal reference mask for the alignment between the layers, and a protection mask used to shield the areas that should not be etched but cannot be covered by the mask created with the e-beam writer.

1.12   PATTERN METAL REFERENCE LAYER

1. Clean the substrate. Spray acetone, then isopropyl alcohol (IPA), and then deionized (DI) water. Repeat the cycle twice or until satisfied with the level of substrate cleanliness.
2. Do a vapor deposition of a thin layer of HMDS. This layer will enhance adhesion of the photoresist to the fused silica substrate. HMDS will adhere to any substrate that has oxygen atoms, including oxide layers (e.g., silicon substrates), but will not help with substrates like GaAs that do not have oxygen.
3. Spin photoresist to a thickness of 1.3–1.4 µm, in this case AZ-5214, and bake it at 90°C for 90 s. The photoresist should be thick enough to allow straight or slightly inclined walls toward the outside to avoid any metal scraps hanging on the sides of the devices. Ensure a highly anisotropic metal deposition. Figure 1.36 shows the three possible profile shapes for liftoff, where the straight walls in Figure 1.36a and the inclined walls toward the inside of the line in Figure 1.36c are undesirable. The walls that are inclined toward the outside of the line as in Figure 1.36b are preferred.
4. Use the contact aligner to pattern the reference layer in the photoresist using the mask designed for the reference metal/fiducials.
5. Apply the developer for 15 s, in this case AZ400K.
6. Rinse with DI water to halt development.

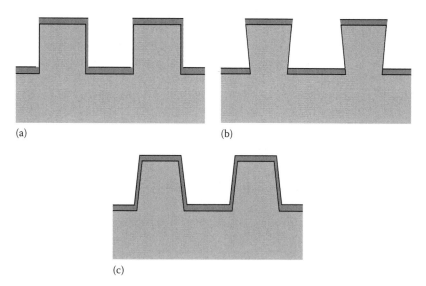

(a)   (b)

(c)

FIGURE 1.36   Three possible grating profiles for liftoff. (a) Straight walls where there is a risk of some metal hanging over the edge. (b) Inclined walls toward the outside of the line where the anisotropy of the deposition should prevent any hanging metal (ideal). (c) Inclined walls toward the inside of the line where there is potential for metal hanging on the sides of the devices.

7. Place the substrate in the ozone (O_3) plasma for 2 min to remove any residual photoresist.

8. Inspect the photoresist pattern using a microscope. If the pattern looks substandard, rework the substrate by starting at step 1.

9. Place a substrate inside a high vacuum deposition chamber ($\sim 5 \times 10^{-7}$ PSI) and deposit 100 Å of Ti or Cr, and 1000 Å of Au. The Ti or the Cr layers are used as adhesion layers between fused silica and Au.

10. Bring substrate from vacuum chamber and perform liftoff. Soak the substrate in acetone for 3–4 h when the open areas between the gold devices are small. This can take up to 24 h if the areas to be lifted are large. The acetone will not penetrate Au/Ti or Au/Cr and can only dissolve the photoresist from the feature edges.

11. Remove any leftover material on the top of the substrate by spraying acetone over the part.

12. Rinse with IPA and DI water.

13. Dry with pressured dry air.

14. Ash for 15 min to remove any residual organic material (photoresist). The ash chamber is used to remove organic materials (e.g., photoresist) with an O_2 plasma.

15. Inspect the metal reference pattern using a microscope. If the pattern looks substandard, rework the substrate by starting at step 1.

1.13 PATTERN AND ETCH FIRST ETCH MASK

16. Spin PMMA to more than 3000 Å (5000 rpm—4% PMMA/96% solvent (chlorobenzene)). The desired feature sizes and the layer etch depth (consequently the thickness of the Ni layer) determine the PMMA and its thickness.

17. Bake PMMA to 170°C for 15–20 min. The process has a 5 min window.

18. Do a high vacuum thermal deposition of 100 Å of Au on top of the PMMA. Using an ion beam-assisted deposition for this step may unintentionally cross-link the PMMA. The Au layer is used to avoid charging on the dielectric substrate.

19. The e-beam writing was done with a 50 kV machine. The lens is radially symmetric so that the feature sizes change across the radius of the device. Smaller features on the outer rims of the lens require higher doses than the larger features in the center of the lens. A dose equals the current multiplied by the time, divided by the area. In the second and third masks the PMMA thickness is higher and the dose is increased, however the features are also larger so the resulting dose is not linear with the thickness.

20. Dip in potassium iodide/iodine (KI) and spray acetone to remove Au from the top of PMMA.

21. Develop PMMA using a 1:3 ratio of methyl isobutyl ketone (MIBK) to IPA for 70 s.

22. The desired profile is shown in Figure 1.36b.

23. Do a high vacuum deposition of $50\,\text{Å}$ of Cr and $500\,\text{Å}$ of Ni.
24. Soak in acetone from 3 h up to 24 h. The patterned metal is going to be the etch mask.
25. Remove any leftover material on the top of the substrate by spraying acetone over the part.
26. Rinse with IPA and DI water.
27. Dry with pressured dry air.
28. Do a SEM and a microscope analysis of the part before proceeding with the etch, ash to clean any organics. If the pattern is substandard, then return to step 16.
29. If there is a need for ash, it should take around 5 min to clean any residual PMMA.
30. Because the features to be etched cover a small area compared to the total substrate area, it is common to protect the area outside the devices from etching. Without protection, those areas will etch and consume the etchant chemicals, causing loading effects that change etch selectivity and uniformity across the substrate. One way to reduce the loading effect is by masking these areas with a protective layer of photoresist. Some thickness of the photoresist will be removed by sputtering and with the reaction of the etchant chemicals. The photoresist should be patterned on top of the substrate using steps 1 through 8 but using a contact mask designed for protection. The photoresist must be thick enough to withstand the etching removal process. It is possible to apply a photoresist layer thickness of 3–4 μm since the features in this protection mask are usually large.
31. Most substrates used for optical purposes cannot have a stop-etch layer as it introduces an undesirable optical interface in the transmission. Consequently, it is necessary to predetermine the substrate etch-rate and etch the component for a prescribed time. Etch rates are functions of chemistry, pressure, bias, the substrate material, the total area to be etched, and the RIE condition. To reduce the variables affecting etch-rate, it is not uncommon to scrub, chemical clean, and condition the RIE to bring it to a well-known starting state.
32. For this case, the RIE is set to a pressure of 40 mTorr, the RF to 200 W, and the DC bias to 434 V. The chemistry is set to 40 sccm of CHF_3 and 3 sccm of O_2. With these parameters, the RIE is expected to etch 0.436 μm in 15–30 min.
33. If there is uncertainty in the etch rates, it is possible to stop at a safe time interval, 80% or 90% of the estimated etch time, bring out the substrate to measure the etch rate, utilizing that parameter to adjust the etch time for the last 20% or 10% of the process.
34. Bring the part out from the RIE, clear one of the depth fiducials with an acetone swab, and measure the depth with a profilometer. If the device is under-etched it is possible to put the part in the chamber for another etch run with updated etch rates and time in the chamber.
35. At the end of the etch, dip the substrate in hydrochloric (HCl) acid for 3–4 min to remove the metal mask.

36. Rinse with DI water to remove the acid. Inspect for leftover metal mask and repeat 35 if necessary.
37. Do a solvent clean with acetone and IPA to remove the photoresist.
38. Put the part in the ash oven for 15 min to remove any leftover photoresist.
39. Inspect the surface of the substrate before proceeding. If the part is out of specifications or there are big errors the substrate must be scrapped.

1.14 PATTERN AND ETCH SECOND ETCH MASK

40. The second etch follows the same steps as the first etch but there are modifications due to a deeper etch of 0.873 μm.
41. Spin PMMA to more than 6000 Å (2000–3000 rpm—6% PMMA/94% chlorobenzene).
42. Develop the PMMA using a 1:3 ratio of MIBK to IPA for 90 s.
43. Double the thickness of Ni to be deposited to 750 Å with the same 50 Å adhesion layer of Cr.
44. Etch time can easily reach 60 min.

1.15 PATTERN AND ETCH THIRD ETCH MASK

45. The third etch follows the same steps as the first etch but there are modifications due to an even deeper etch of 1.745 μm.
46. Spin PMMA to more than 12000 Å (2000–3000 rpm—9% PMMA/91% chlorobenzene).
47. Develop the PMMA using a 1:2–1:3 ratio of MIBK to IPA for 120–180 s.
48. Increase the thickness of Ni to be deposited to 1000 Å with the same 50 Å adhesion layer of Cr.
49. Etch time can easily reach 120 min.

ACKNOWLEDGMENTS

The authors wish to acknowledge the following collaborators: J.R. Wendt, D.W. Peters, R.R. Boye all at Sandia National Laboratories, Albuquerque, NM, 87185, USA; T.R. Carter at Sandia Staffing Alliance, Albuquerque, NM, 87110, USA; S. Samora at LMATA, Albuquerque, NM, 87109, USA; S.H. Zaidi at Gratings Inc., Albuquerque, NM, 87107, USA; and D.S. Ruby.

Sandia is a multiprogram laboratory operated by Sandia Corporation, a Lockheed Martin Company, for the United States Department of Energy's National Nuclear Security Administration under contract DE-AC04-94AL85000.

REFERENCES

1. W.L. Wolfe, ed., *Optical Engineer's Desk Reference*, Optical Society of America, Washington, D.C., Chapter 17, 2003, pp. 347–370.
2. G.J. Swanson, Binary optics technology: The theory and design of multilayer diffractive optical elements, MIT Lincoln Laboratory Tech. Rep. 854, NTIS Pub. AD-A213-404.

3. G.A. Vawter, Ion beam etching of compound semiconductors, in *Handbook of Advanced Plasma Processing Techniques*, R.J. Shul and S.J. Pearton (eds.), Springer, Berlin, 2000, pp. 507–547.

4. S.A. Kemme, J.R. Wendt, G.A. Vawter, A.A. Cruz-Cabrera, D.W. Peters, R.R. Boye, C.R. Alford, T.R. Carter, and S. Samora, Fabrication issues for a chirped, subwavelength form-birefringent polarization splitter, *Proceedings of the SPIE—The International Society for Optical Engineering*, 6110, 112–119 (2006).

5. R.R. Boye, S.A. Kemme, J.R. Wendt, A.A. Cruz-Cabrera, G.A. Vawter, C.R. Alford, T.R. Carter, and S. Samora, Fabrication and measurement of wideband achromatic waveplates for the mid-infrared region using subwavelength features, *Journal of Microlithography, Microfabrication, and Microsystems*, 5, 043007, October–December (2006).

6. S.A. Kemme, D.W. Peters, T.R. Carter, S. Samora, D.S. Ruby, and S.H. Zaidi, Inadvertent and intentional subwavelength surface texture on microoptical components, *Proceedings of the SPIE—The International Society for Optical Engineering*, 5347, 247–254 (2004).

7. Valley Design Corp., Santa Cruz, CA.

8. P.B. Claphan and M.C. Hutley, Reduction of lens reflection by the 'moth eye' principle, *Nature*, 244, 281–282 (1973).

9. M.E. Motamedi, W.H. Southwell, and W.J. Gunning, Antireflection surfaces in silicon using binary optics technology, *Applied Optics*, 31, 4371–4376 (1992).

10. D.S. Ruby, S.H. Zaidi, and S. Narayanan, Plasma-texturization for multicrystalline silicon solar cells, *Twenty-Eighth IEEE PVSC*, pp. 75–78 (2000).

11. D.S. Ruby, S.H. Zaidi, B.M. Damiani, and A. Rohatgi, RIE-texturing of multicrystalline silicon solar cells, *PVSEC-12*, Jeju, Korea, pp. 273–274, June 2001.

2 Fabrication of Microoptics with Plasma Etching Techniques

Gregg T. Borek

CONTENTS

2.1 INTRODUCTION AND OVERVIEW

This chapter presents and reviews etching techniques for microoptics fabrication that were developed at both MEMS Optical, Inc. and related companies during the past 16 years. Many practical examples and applications are presented. The principal method of fabricating microoptics at MEMS Optical has been using gray-scale photolithography for patterning and plasma etching for transferring the pattern into a substrate. The gray-scale manufacturing technique has allowed fabrication of a wide range of microoptical devices for use in a broad range of applications. The photolithographic patterning method and etch pattern transferal techniques can be used in the manufacture of both diffractive and refractive microoptical devices. Some of the more commonly manufactured microoptical elements include beam splitting diffractive optics, beam shaping diffractive optics, diffusing or homogenizing diffractive optics, diffractive lenses and lens arrays, refractive microlenses and microlens arrays (MLAs), and other phase-modulating optics. The functional microoptics have been produced in a wide range of substrate materials for addressing applications across the electromagnetic spectrum. Microoptics have been fabricated for use across the optical radiation spectrum that ranges from deep ultraviolet (DUV) at the wavelength of 157 nm through long-wave infrared (LWIR) at the wavelength of 14 μm. Microoptics have been fabricated into many substrate materials that have transmission bands within the optical spectrum including fused silica, silicon, germanium, gallium phosphide (GaP), gallium arsenide (GaAs), gallium nitride (GaN), silicon carbide, Pyrex, borosilicate glasses, high refractive index optical glasses, flat panel display glasses such as C-1737 and Eagle2000, zinc selenide (ZnSe), multispectral zinc sulfide (ZnS), sapphire, and calcium fluoride. Some of the high index glasses and materials investigated were OHARA S-TIH-53, OHARA S-LAH79, and Schott SF57. The chapter describes the fundamental steps in manufacturing microoptics with an overview of the gray-scale photolithography required to produce the three-dimensionally patterned microoptic structures in a variety of material systems.

Gray-scale photolithography uses the technique of creating a photomask that spatially alters the intensity of optical radiation transmitted through the photomask. Several methods have been employed in performing this task including using photomasks that have a spatially variable optical density,[1,2] using photomasks that have a spatial variation in optical absorption,[3] and masks that use variable-sized microapertures to modulate the amount of light transmitted locally in the mask field. Gal[4] and others developed one of the first technologies of gray-scale photolithography and pattern transfer techniques to manufacture microoptical elements.[5] The gray-scale fabrication technique uses a single photolithography iteration to three dimensionally pattern a photosensitive polymer with the desired microoptic structure.

The ability to use gray-scale fabrication allows the optical designer much freedom in creating microoptical solutions.[6] Positive (convex) surfaces and negative (concave) structures can be combined in the same optical element. The photoresist shaping from gray scale enables the fabrication of anamorphic or saddle lenses having positive and negative curvatures along orthogonal axes. The saddle-shaped refractive surface can be formed for the wave front correction of astigmatic sources. Aspherical lenses as well as parabolic cylinder lenses can be readily fabricated through the use of gray-scale manufacturing techniques. Microoptical elements that perform multiple functions on one optical surface such as focusing light and splitting light into multiple beams can be realized. Nontraditional optical elements include the phase surface microoptical components such as phase diffusers, beam integrators, and gratings or spot generators and have been produced with gray-scale photolithographic techniques.

Some common examples of microoptical structures fabricated with gray-scale techniques are shown in Figure 2.1. The examples include a smooth kinoform diffractive lens, a dual functionality microoptical element with focusing and beam splitting occurring at the same optical surface, and high fill factor positive and negative MLAs. The photolithographic patterned features can be permanently transferred onto the substrate surface using reactive ion etching (RIE) techniques. The reactive

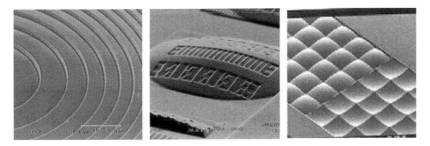

FIGURE 2.1   SEM images of typical microoptical elements fabricated with gray-scale processing including diffractive lens, dual functioning focus and beam splitting lens, and convex and concave MLA.

etch pattern transfer method can use capacitively coupled etch tools or high-density plasma tools, including inductively coupled plasma (ICP) etching. Plasma etching techniques established and perfected in the microelectronics industry have been modified and adapted to successfully etch wafers containing functional microoptical devices. The industry standard plasma etching processes for oxides, silicon, and compound semiconductors have been enhanced and adjusted to work with microoptical structures that are transferred into these various substrate materials.

The ability to fabricate microoptical structures is affected by a combination of both the photolithography process and by the pattern transferal or etching process. The selectivity of the etching transferal process is one of the key parameters for producing microoptics and relates the pattern height formed in the photosensitive polymer to the pattern profile height after transferal into the substrate material. The three-dimensional (3D) structure formed in the photosensitive polymer serves as a sacrificial etching mask layer. After the etch transferal, the mask layer is completely etched or consumed by the process. Selectivity is defined as the ratio of the material etch rates of the substrate material versus the etch mask material. The maximum achievable profile height or lens sag height may be limited by a combination of the photolithography and etching process. The selectivity properties can be used to stretch or compress the topographical height of the sacrificial etching mask layer.

Reactive ion etch (RIE) processing detailed in this chapter was performed in modern high density plasma etch tools. The etching machines are of the ICP architecture.[7] ICP etching tools employ two radio frequency (RF) generators to produce the reactive ions in the plasma that are required for etching substrate materials. One RF source drives the coil in the upper part of the processing chamber to ionize the etch gas and generate the high density plasma. The RF generator frequency can typically be 13.56 or 2 MHz, depending on the RF impedance matching network design. The second RF generator, typically at 13.56 MHz, drives the platen electrode and applies a bias to the wafer. This RF configuration allows generation of a high ion density while allowing for independent control of the bias voltage applied to the wafer. The ion density in the ICP architecture is typically an order of magnitude greater than the ion density of a capacitively coupled RIE, so the ICP substrate etching rates can be much greater than for standard RIE processing. This etch environment affects the wafer or substrate being processed and requires an effective substrate cooling system

FIGURE 2.2　STS oxide etcher (left), STS silicon etcher (center), SLR770 etcher (right).

to protect the photoresist-based etch mask material on the substrate surface. These systems are equipped with direct wafer contact bottom side helium cooling that can be regulated and controlled to provide uniform and stable cooling for the wafer during etch processing. To maintain helium pressure on the bottom surface of the wafer or substrate, the wafers are constrained to the platen with a clamping technique. Several clamping implementations that have been integrated into the processing machines use mechanical clamping, weighted clamping, or electrostatic clamping.

The tools have the capacity to run many different processing gases for reactively etching different substrate material systems and for controlling etch quality. The etch tools are generally equipped with high conductance corrosive service turbo-molecular pumps rated from 1000 to 2000 L/s pumping speeds. This pump feature allows the processing to support high etch gas mass flows while maintaining a low chamber pressure. The mechanical backing pumps are rated for corrosive service and for maintaining low foreline pressure for high gas flows. Typical ICP chamber pressure range for processing is from 1.5 to 30 mTorr. Figure 2.2 shows the etching tools used for processing these various wafer materials. For compound semiconductor processing, the SLR770 is used. The STS oxide etcher is used for glass systems, and the STS ASE® etcher for silicon materials.

2.2　ETCH PROCESSING CONSIDERATIONS

The functional result of plasma etching is to create volatile compounds by forming a reaction between the ionized gases in the plasma and the substrate surface so that the volatile compounds can be easily pumped out of the etching system. The etching process must also react with and etch or remove the sacrificial polymer-based etch mask. For gray-scale-based microoptical device processing, as the etching process progresses, the etch mask is entirely removed. Volatile compounds are pumped out of the processing chamber. Volatile species are created for both chlorine and fluorine-based processing chemistries. Plasma etch processing should produce an etched wafer that is free from pitting and redeposition of the etch mask and substrate. Pitting can either occur in the wafer material, or in the etch mask material that is proportionally transferred into the wafer by the etch. Redeposition of reacted products can produce micromasking or cause micrograss formation to occur. Micrograss is characterized by tall narrow structures, resembling grass, that are formed by the redeposition of etch material that acts as a localized mask on the micro- or nano-scale. Figure 2.3 shows an example of micromasking encountered during GaP etch development. Figure 2.4 illustrates results of an improved etching process to eliminate the etched by-product redeposition. The observed quality improvement was the result of varying the mixture ratio of the etch gases. Throughout the process, the mask material is gradually removed for gray-scale microlens fabrication and the etching process must be stable for the duration of the pattern transfer. The etch process must maintain the quality of the photoresist lens or photoresist surface during the pattern transferal into the wafer.

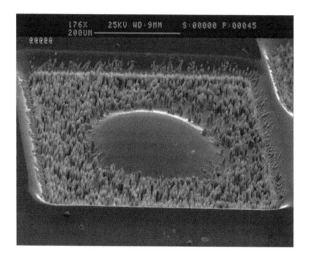

FIGURE 2.3   Severe etch process micromasking.

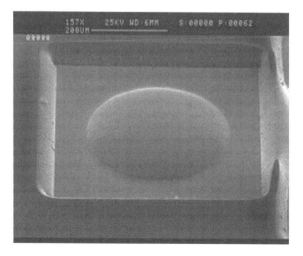

FIGURE 2.4   Improved etch process with no micromasking.

Selectivity is a very significant processing parameter that considers both the masking material and the substrate material. Selectivity is defined as the ratio of the etch rates of the substrate material to the masking material. For selectivity greater than 1, the wafer etches more rapidly than the masking material. The selectivity for a photoresist and wafer material can limit the maximum profile depth achievable. For quality processing results, the etching process selectivity must be repeatable from run to run over a large number of wafers. For example, in an unstable etch process, processing by-products such as polymer films can redeposit on exposed surfaces and hardware inside the processing chamber. The redeposition or condensation of reacted materials can affect etching repeatability by making the process selectivity unrepeatable. The etch process should be predictable to adjust for slight surface height variations from photolithography processing. For some etch processes, a plasma chamber cleaning needs to be run after every wafer or every few wafers to maintain process repeatability. Finally, for production considerations, the process should also be capable of etching a sufficient number of wafers without having to perform an invasive chamber cleaning procedure. The invasive cleaning includes opening the chamber for mechanical scrubbing or wet cleaning, followed by a conditioning sequence.

A selectivity curve versus a process adjustment parameter can be determined through experimentation to define selectivity behavior versus the variation of a parameter or parameters. A design of experiments (DOE) investigation can be performed to determine etching parameters or variables significant to the process for etching the substrates. DOE software can be used to analyze the experiment to predict the variables influence on the process and model the interactions between variables of the etch process. When performing an etch development effort on a new wafer material, the material system can be researched for similarities with other materials to find acceptable starting conditions from published research references. DOE is a powerful analytical method that can then be executed to define the effects of processing parameters.

The modern ICP etching tools have a large number of degrees of freedom for defining and controlling etching processes. The following parameters have been used to define selectivity curves for various materials: coil RF power, platen RF power, process chamber pressure, etch gas species content, concentration of gas species, dilution gas species content, wafer temperature, and equipment temperature. These parameters can also be varied over time for interesting photoresist shaping effects. Hence, many tool parameters or variables have an effect on plasma etching processes. Figures 2.5 through 2.7 depict plots of selectivity versus a processing parameter for three different types of etch processing for the material systems being considered—glass, silicon, and GaAs—that were processed at MEMS Optical.

In standard etch processes for glass materials such as fused silica, quartz, and Pyrex, selectivity is controlled by the addition of oxygen as a processing gas. For some photoresist patterns, the oxygen content can cause a rough or pitted surface. An alternative method of selectivity control is depicted in Figure 2.5. The ratio of two fluorocarbon gases can be varied while keeping the total mass flow constant to vary selectivity in a predictable fashion. The ratio modifies the relative of amount of CF_x precursors that determine how much photoresist etching mask relative to the glass substrate is etched. Figure 2.6 depicts a selectivity control curve for silicon

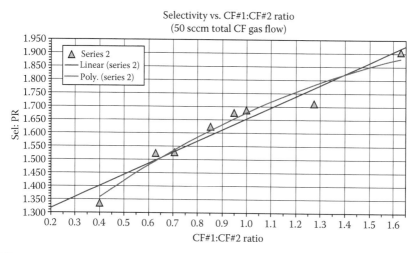

FIGURE 2.5   Selectivity curve for fluorocarbon-based glass etching.

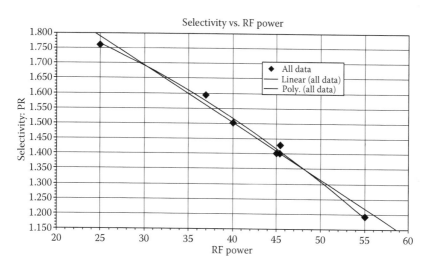

FIGURE 2.6   Selectivity curve for fluorine-based silicon etching.

etching by varying the amount of RF power supplied to the platen and allowing the etch chemistry to remain unchanged. The reduced RF power minimizes etch mask removal relative to silicon. This curve illustrates the ability to predictably make minor adjustments for selectivity, for example, adjusting to within a percentage of a targeted selectivity.

Coarse adjustment of selectivity for a chlorine chemistry material process is depicted in Figure 2.7. The selectivity of GaAs can be adjusted by changing the ratio of chlorine to boron trichloride processing gas while maintaining the total mass flow constant. The curve establishes a general trend of increasing selectivity with increasing $Cl_2:BCl_3$ for a wide range of chamber pressures and bias voltages.

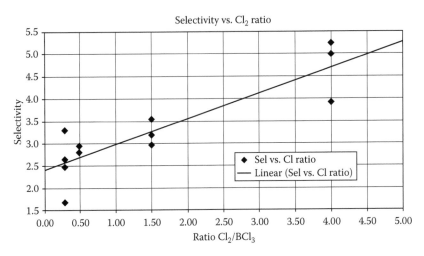

FIGURE 2.7   Selectivity curve for chlorine-based GaAs etching.

The multiple data points along a given gas ratio indicates varied operating set points for other parameters. The coarse curve could be used to set the general selectivity, and a different parameter could define a fine adjustment selectivity curve.

2.3   ETCH PROCESSING GLASS SYSTEMS

Several optical glass materials were investigated for etch processing quality for microlens applications.[8] The glasses considered have indices of refraction that are double the refractive index for fused silica. The materials were experimented for manufacturing high index of refraction material hemispherical lenses in wafer form. The materials considered are the Ohara glasses S-TIH-53 and S-LAH79 and Schott glass SF57. The indices of refraction for the deep blue wavelength ($\lambda = 405$ nm) are 2.07 for S-LAH79, 1.91 for S-TIH53, and 1.91 for SF57. The materials transmit in the wave band range of 390–2400 nm. The optical parameters referenced here were listed on the manufacturer Web sites. The other glass considered is Corning C1737, an alkaline earth boroaluminosilicate. C1737 and other derivative glass is widely used in LCD flat panel display applications and is available in relatively inexpensive sheet form. The C1737 is an inexpensive alternative to fused silica for low cost glass microoptical applications. Etching optical structures into this material enables structures to be built and assembled into the Active Matrix Liquid Crystal Display (AMLCD) stack. Thermal coefficient of expansion (TCE) differences can be avoided for mismatched materials. It is difficult to obtain the exact elemental composition of the glasses, however, it is believed that the TIH glasses have titanium oxides, the LAH glasses are lanthanum compounds, and SF glasses contain lead oxides. Figure 2.8 depicts a MLA with individually rotated and tiled microlenses for a 3D display application. The vertex of the clipped rectangular lenslet runs along the diagonal axis of the clipped lens aperture. The MLA was fabricated in Corning Eagle2000 display glass. The lens parameters include 6.0 µm center to edge sag height, with lenslet dimensions of 94 µm wide × 340 µm tall, and a radius of curvature (RoC) 190 µm. The etch process selectivity was 1:1 with glass etch rate 1500 Å/min.

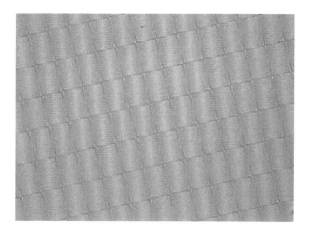

FIGURE 2.8 Rotated and clipped MLA for stereoscopic (3D) display application.

The optical glasses are more challenging to reactively etch in a plasma process than fused silica and Pyrex because they contain elements and compounds that do not react in the etch chamber to form volatile fluoride compounds. The etch processes for these optical glasses are mechanically dominated or closer to a sputter process and require high bias voltages to etch. Hence, these aggressive processes tend to cause photoresist etch mask degradation that results in lower process selectivity. Because of the other nonvolatile compounds present, the etch rates tend to be lower than for fused silica. The maximum level of high bias voltages is limited by photoresist reticulation caused by heating the substrate during processing. Table 2.1 provides the tabulation of etching results for processing these optical glass materials. Parameters are compared for the optical glasses versus fused silica. The parameters listed are maximum and minimum selectivity to photoresist, maximum and minimum etch rate for the glass material, and maximum and minimum bias voltage. The bias voltage is an indication of the aggressiveness of the etch process. The high bias voltage can create a level of ion bombardment that causes significant wafer heating. Out of the glasses considered, SF57 etches most similarly to fused silica. The table is not a complete operating range of all possible combinations of etch parameters, merely a presentation of results obtained through processing these materials at MEMS Optical. The materials with selectivity less than 1 are not as useful for producing microoptics as materials

TABLE 2.1

Etching Results for Optical Glass Materials

Etch Property	Fused Silica	S-LAH79	S-TIH53	SF57	C1737
Maximum selectivity	3.8:1	0.63:1	0.84:1	3.2:1	1.68:1
Minimum selectivity	0.3:1	0.2:1	0.18:1	2.2:1	0.3:1
Maximum etch rate (Å/min)	4000	625	1020	3550	1725
Minimum etch rate (Å/min)	1100	170	216	2550	570
Maximum bias voltage (V)	325	750	508	480	700
Minimum bias voltage (V)	225	200	226	360	310

with higher selectivity. The lenses were patterned using gray-scale processing with positive photoresist. Possibly other mask agents can yield a higher selectivity process. Soft lithography processing or replication of other mask agents could improve selectivity by increasing the etch resistance of the lens etch mask. Employing other masking materials remains to be investigated. Sag depths attainable for the various materials can be limited by both the etch rate and selectivity. Etch rate determines exposure duration of the mask material to the plasma conditions.

Table 2.2 shows the before etch and post etch difference in surface roughness due to processing. The peak-to-valley (PV) surface roughness data presented are for a $140 \times 140\,\mu m$ area. The root mean square (RMS) surface roughness is a better representation of the overall surface quality. Surface pits or spikes tend to skew the PV calculation. Figure 2.9a shows a micrograph of the gray-scale lens that is etched into the SLAH79 wafer. The flat top at the lens vertex indicates that the etching process was stopped prematurely, before the photoresist mask pattern was fully transferred into the wafer. Figure 2.9b is a NewView 5000 solid scan plot view of the lens shape. Figure 2.9c shows the lens surface quality with the lens shape subtracted from the data. The PV is 110 Å with an RMS roughness of 12 Å. The gray-scale lens etched into the SF57 glass wafer is shown in Figure 2.10a through c. Figure 2.10a shows a micrograph image of the $360\,\mu m$ diameter lens that is etched into the SF57 wafer. Figure 2.10b shows a zoomed in ZYGO solid image of the etched surface quality.

TABLE 2.2
Etching Results for Surface Quality for Optical Glass Materials

Material	Data Source	Peak-to-Valley (Å)	RMS Roughness (Å)	Sag Etched (μm)
S-LAH79 3 in. diameter	Wafer surface	42	7	
	Lens surface	110	12	4
S-TIH53 4 in. diameter	Wafer surface	177	13	
	Lens surface	480	82	18
SF57 4 in. diameter	Wafer surface	160	16.6	
	Lens surface	800	54	10.5
C1737 3 in. diameter	Wafer surface	110	10.4	
	Lens surface	400	167	5

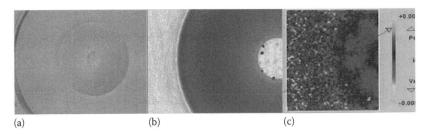

(a)   (b)   (c)

FIGURE 2.9   (a) Micrograph image, (b) ZYGO solid plot, and (c) surface RMS after etching $4\,\mu m$ sag lenses into the SLAH79 wafer surface.

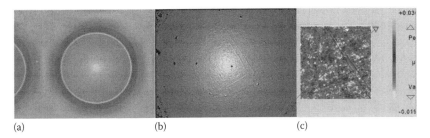

(a)   (b)   (c)

FIGURE 2.10   (a) Micrograph image, (b) ZYGO solid plot, and (c) surface RMS after etching 10.5 µm sag lenses into the SF57 wafer surface.

Figure 2.10c shows the lens surface quality after etching with the lens shape subtracted from the data. The PV roughness is 800 Å and the RMS roughness is 53.5 Å.

2.4   ETCHING MICROOPTICAL STRUCTURES IN SILICON

The silicon microoptic and lens etch process has been well characterized over the past 15 years. For silicon microoptics, the plasma etch processing chemistry is the fluorine-based process using SF_6 as well as other C_xF_y gases. The selectivity for using positive photoresist ranges from 0.5:1 to 3.0:1. Silicon is very reactive at low bias voltages, so fairly high etching rates are achievable. From processing various microoptical elements, the following substrate etch rates ranging from 0.25 to 2.0 µm/min range were observed. The processes developed for etching silicon have been used to produce a wide variety of gray-scale patterned structures including axially symmetric aspherical lenses (positive and negative), cylindrical lenses (positive and negative), and biconic lenses. Biconic lenses contain curved surfaces where the RoC values differ in transverse or orthogonal orientations and the conic constant values in orthogonal orientations are not equal. Figure 2.11 depicts a biconic lens used for coupling the output of laser diode into an optical fiber. Figure 2.12 shows the lens in an arrayed format as fabricated on the wafer prior to separation by wafer dicing. Figure 2.13 depicts the manufactured lenses on the back surface of the biconic lenslet, which is depicted in Figures 2.11 and 2.12.

FIGURE 2.11   Biconic silicon lens with RocX/RocY ratio of 5:1 conic constant Kx ≠ Ky.

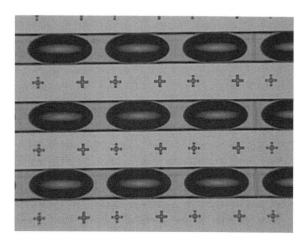

FIGURE 2.12 Fabricated array of biconic lenses before dicing—lens size 600 × 250 µm, with 30.5 µm lens sag.

FIGURE 2.13 Negative cylindrical lens with one-dimensional (1D) asphere prescription, gold solder posts and lithographically defined antireflection coated region.

The lenslet depicted in Figure 2.13 is a negative cylindrical lenslet having a parabolic surface. Two additional features that can be observed are lithographically defined gold solder posts for use in reflow bonding the microlens to the laser diode and a lithographically defined aperture on the surface of the negative lens, so that the antireflection coating only deposited on the lens surface and not onto the surrounding wafer surface.

A similar silicon etch process can be used for manufacturing microoptical gratings for laser wavelength tuning. For the application depicted in Figure 2.14, the grating is a gray-scale sinusoidal grating with a pitch period of 810 nm. The white light reflected by dispersion of the grating is visible in the image. The peaks and valleys comprising the sinusoidal grating are 405 nm wide. To obtain the surface topology, the gratings were measured with an atomic force microscope (AFM). The AFM scan is depicted in Figure 2.15.

FIGURE 2.14 Micrograph images of C-band and L-band tunable gratings for use in the Littrow configuration.

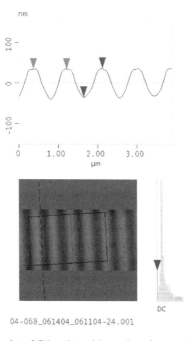

04-068_061404_061104-24.001

FIGURE 2.15 AFM scan plot of C-band tunable gratings for grazing incidence configuration.

2.5 ETCH PROCESSING III–V MATERIALS WITH GRAY-SCALE MICROOPTICAL STRUCTURES

A significant volume of work has been performed on etching gray-scale photolithographic structures into the high index of refraction materials GaP and GaAs. Note that many different combinations of etching parameters are used to generate the range of etching conditions.[9,10] Table 2.3 summarizes the observed performance results for processing these materials at MEMS Optical. Figures 2.16 through 2.19 depict several precision microlens applications using gray scale for both GaP and GaAs. Figure 2.16 shows a micrograph of a combination of low selectivity (0.5:1) etching for an on-axis 25 mm RoC lens and high (3.5:1) selectivity etching for a 0.9 mm RoC off-axis lens as two process steps for creating the microoptical device in GaAs. Figure 2.17 is a micrograph of an off-axis gray-scale diffractive microlens

TABLE 2.3

Etching Parameters for GaP and GaAs

Property	GaP	GaAs
Maximum selectivity	3.5:1	6.0:1
Minimum selectivity	0.6:1	0.37:1
Maximum etch rate	3.1 μm/min	3.5 μm/min
Minimum etch rate	0.62 μm/min	0.31 μm/min
Wafer surface (RMS)	Epitaxial grade 3–10 Å	Epitaxial grade 3–10 Å
Lens etch (RMS)	30–350 Å	30–400 Å
Wafer availability	2 and 3 in. diameter	2 in.–150 mm diameter
Optical transmission	0.58–1.8 μm	0.95–19.0 μm

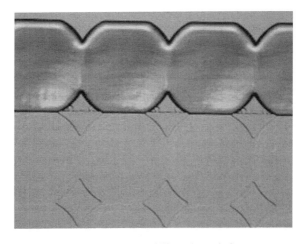

FIGURE 2.16   Dual RoC device in GaAs on 250 μm lens pitch.

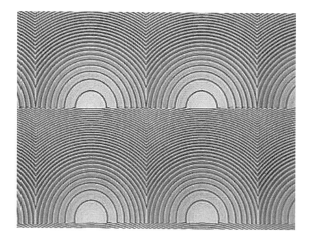

FIGURE 2.17 Off-axis diffractive in GaAs on 250 μm lens pitch.

FIGURE 2.18 On-axis 330 μm diameter aspherical lens in GaP as singlet.

in GaAs. Figure 2.18 is a micrograph of a precision gray-scale microlens 360 μm in diameter with a 30 μm sag height. Figure 2.19 is a ZYGO Newview scan of the lens depicted in Figure 2.18. This ZYGO scan shows the lens surface quality after the lens shape is subtracted. Nominal surface RMS roughness after processing is 22.8 Å. The standard dimension of GaP wafers available from suppliers is 50 mm diameter wafers. Additional examples showing wafer-level fabrication of microlenses in GaP material using gray-scale manufacturing are depicted in Figures 2.20 through 2.23. The applications include off-axis lens array for a telecom coarse wavelength division multiplexing (CWDM) product wafer depicted in Figure 2.20 and an on-axis aspherical lens for fiber collimation for a telecom application shown in Figure 2.21.

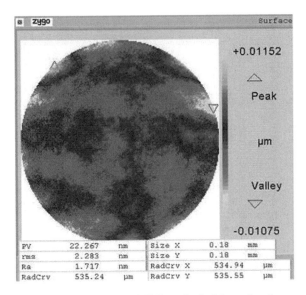

FIGURE 2.19 ZYGO Newview scan of GaP aspherical lens showing post etch surface quality.

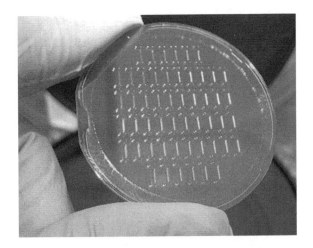

FIGURE 2.20 Linear off-axis aspherical MLA for CWDM application.

Figure 2.22 is a micrograph of truncated gray-scale aspherical lenses in GaP for collimating an array of many arrayed laser diodes. The microlens dimensions are 50 μm wide by 130 μm long, having a lens sag 20 μm tall. Figure 2.23 depicts a wafer with 1 mm by 1 mm dimension lenses for a defense application using a high fill factor Shack–Hartmann MLA for SWIR wave front sensing.

FIGURE 2.21   Microlens with 360 μm diameter aspherical lens for telecom fiber coupling application.

FIGURE 2.22   Wafer of 140 × 50 μm truncated aspherical lens array for arrayed diode laser collimation application.

2.6   ETCHING MICROOPTICS IN OTHER MATERIALS OF INTEREST FOR THE OPTICAL SPECTRUM GaN, SiC, AND Al$_2$O$_3$

The developmental work detailed in this section was performed from year 2001 through early 2002. The materials were investigated as a possible wafer solution for fabricating arrays of high numerical aperture (NA) hemispherical lenses that would function at $\lambda = 405$ nm for future generation optoelectronic devices.[11,12] This information is shown to reference the volume of business opportunity using the GaN wafer material. When the developmental work was performed, it was challenging

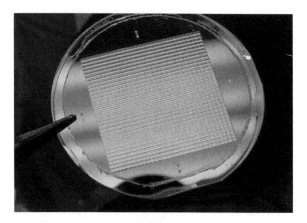

FIGURE 2.23 Wafer of 1 × 1 mm 100% FF MLA for SWIR military application.

FIGURE 2.24 Two inch diameter wafer of GaN on sapphire with etched hexagonal MLA.

to obtain commercially available GaN on sapphire wafers for process testing. The developmental work was performed using 2 in. wafers of MOCVD-deposited GaN on sapphire substrates. The approximate thickness of the GaN layer was 2.1 μm. At the present time, many commercial sources for GaN material exist. Figure 2.24 depicts microlenses etched into the surface of a GaN on sapphire wafer. GaN has an approximate index of refraction of 2.4 for blue wavelengths.[13]

A chlorine-based etch chemistry was used to etch GaN with a selectivity range of 0.81:1 to 0.95:1. The etch rates achieved ranged from 4130 to 8050 Å/min. The restricted availability of wafer etch samples limited the extent of etching development. Figures 2.25 and 2.26 are Newview scans showing the pre-etch wafer surface quality. The PV surface roughness of Figure 2.25 for unetched GaN is 790 Å with an RMS surface roughness of 124 Å. Figure 2.26 depicts the surface texture prior to etching. The post etch surface quality of GaN after the lens shape was subtracted from the

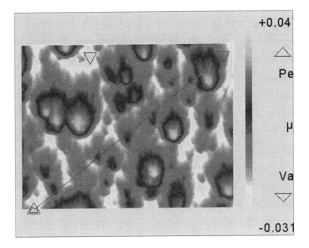

FIGURE 2.25   Newview scan of unetched GaN surface.

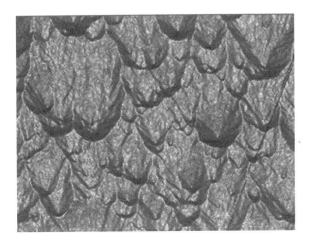

FIGURE 2.26   Newview solid plot of GaN surface texture in Figure 2.25.

resulting lens profile is depicted in Figures 2.27 and 2.28. Figure 2.27 shows the residual surface after the lens shape was subtracted from the measured lens surface. Figure 2.28 is a solid plot showing the residual surface texture. The physical nature of the etch process appears to have smoothed the surface roughness of the unprocessed wafer. The PV roughness of Figure 2.27 for etched GaN is 315 Å with an RMS roughness of 29 Å. The lens shape data was subtracted from the measurement data. The lens sag height was 1.8 µm and was etched into a 2.1 µm thin film layer of GaN. This process illustrates the possibility of including microoptics on active device wafers for optoelectronic applications.

Silicon carbide was selected as a candidate optical material for etching for its high index of refraction in the blue wave band. The substrate material tested for

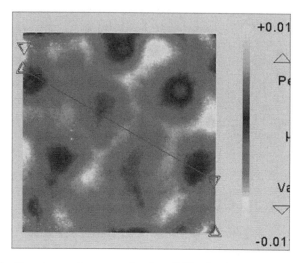

FIGURE 2.27 Newview surface scan showing GaN microlens surface micro roughness after best fit lens shape subtraction.

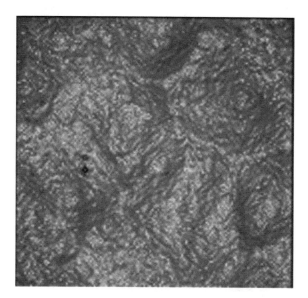

FIGURE 2.28 Newview solid plot of etched GaN shown in Figure 2.27.

etching characteristics is provided by Sullivan Datanite®. Datanite[14] is a specially grown form of 3-C oriented polycrystalline semi-insulating SiC material. Datanite is unique from other SiC crystalline forms because it has no micropipe defects. The material is partially transmitting in the visible, and has a grayish brownish color. This 2 in. diameter Datanite material was obtained for etch testing because crystalline orientation 4H-n wafer was unavailable. The 2 in. samples cost approximately $2000 per wafer. Figure 2.29 depicts a quartered section of a 2 in. diameter SiC wafer with microlenses etched into the surface. The wafers are a very hard material that is difficult to cleave or break. For the wafers to be separated into several etch

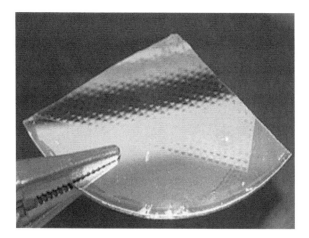

FIGURE 2.29   Two inch diameter Datanite with etched MLA.

process samples, the wafers were repeatedly scribed with a diamond-tipped scribe to form a score line for cleaving. The sample wafers were not able to be cleaved on the scribe lines. The SiC wafers were separated into samples by banging the back end of the scribe tool into the back side of the patterned wafers. The wafers fractured into approximately 4 or 5 pieces per 2 in. wafer. Figure 2.30 shows the index of refraction versus wavelength[15] for the 4H orientation of SiC. For blue wavelength applications, 4H–SiC has an apparent refractive index greater than 2.7. After etch testing was completed on the Datanite SiC, some 4H-*n* SiC material was received. The 4H-*n* material from Bandgap Technologies appears as water clear, with good transmission in the visible. Etch testing was performed on the 4H material with similar results.

Figure 2.31 is a micrograph of the MLA with DIC filtering. The lenses are 800 μm in diameter and are etched to a sag height of 2.5 μm. Due to the limited number of wafer samples, the quartered samples were attached to a carrier wafer with thermally

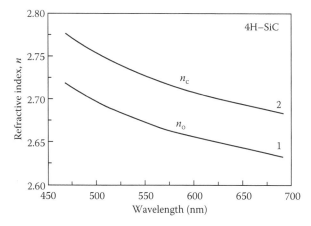

FIGURE 2.30   Index of refraction curve for 4H–SiC. (From world wide web link: http://www. ioffe.ru/SVA/NSM/Semicond/SiC/optic.html; Shaffer, P.T.B., *Appl. Opt.*, 10, 1034, 1971.)

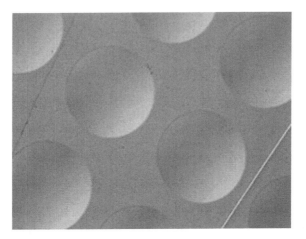

FIGURE 2.31   SiC lenses 800 μm diameter etched in Datanite.

conductive vacuum grease. The etching investigation was performed using a DOE study with a limited scope of experimental variables because there were a limited number of etch samples. The DOE type was a three variable with full factorial, so 12 etch runs were performed to include repeated samples and midpoints. The selectivity for SiC to photoresist did not vary significantly, ranging from 0.53:1 to 0.60:1. The silicon carbide etch rate ranged from 2000 to 4000 Å/min.[16–19] These material etch rates for SiC are similar to material etch rates for fused silica. The etch test produced an acceptable lens surface and lens shape quality. Both surface roughness and process selectivity could be improved with selectivity approaching a goal of 1:1 selectivity ratio. The processing used the sulfur hexafluoride-based process.

The views in Figures 2.32 through 2.34 were from lens scans obtained with the ZYGO NewView 5000. The figures depict a solid view of the lens scan in

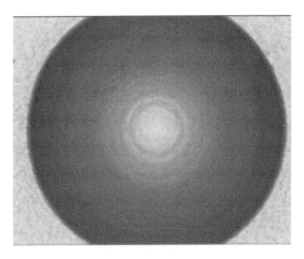

FIGURE 2.32   Solid plot of etched SiC microlens.

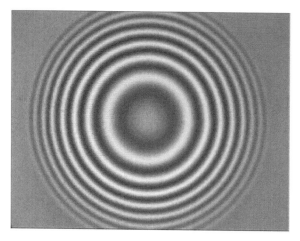

FIGURE 2.33   Newview white light fringe plot of SiC microlens.

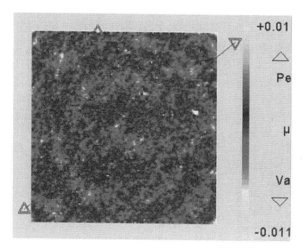

FIGURE 2.34   Newview plot of RMS surface micro roughness after subtracting best fit lens shape from 2.5 μm sag lenses etched in SiC.

Figure 2.32, the white light fringe pattern from the lens surface in Figure 2.33, and the lens surface quality after the lens shape is subtracted out from the profile scan in Figure 2.34. The PV roughness for unprocessed SiC is 220 Å with an RMS roughness of 21.7 Å. From Figure 2.34, the processed silicon carbide surface quality shows PV roughness of 325 Å and an RMS roughness of 21.0 Å. The lens etching process did not significantly degrade the surface quality of the unprocessed SiC wafer. The processing performed demonstrates the feasibility of etching microoptical elements directly into wafers of GaN or silicon carbide. Microoptical elements may be etched with active devices fabricated in these materials.

Etch process development was performed for sapphire (Al_2O_3) crystalline material for microoptical applications.[20–22] Etch processes were developed for C-plane

FIGURE 2.35   Sapphire 7 by 7 beam splitter grating.

and R-plane substrate slices. The spectral transmission window ranges from 200 nm in the DUV through 5.5 μm in the mid-wave infrared region (MWIR). The material, due to its mechanical properties, offers a good optical solution for high power applications in the near infrared (NIR) through MWIR spectral region. Microoptics have been produced on various wafer diameters from 50 to 125 mm in diameter. Microlenses as well as diffractive phase optics such as beam splitters have been produced using the gray-scale fabrication methods. Etch processing uses the chlorine-based chemistry to achieve etch rates similar to fused silica. One potential drawback is the usable selectivity range, which could limit fabrication to the maximum structure height provided by the combination of lithography and etching process. Presently, for the gray-scale process, the maximum selectivity achieved is 0.9:1 for etch mask to sapphire. The maximum sag height for lenses produced has been 18 μm tall. For a diffractive phase grating example, Figure 2.35 shows a sapphire beam splitter grating for dividing an input beam into an output pattern of 7 by 7 output beams. The application for the depicted diffractive optic is laser hole drilling.

2.7 ETCH PROCESSING II–VI MATERIALS ZNSE AND MULTISPECTRAL ZNS

ZnSe and ZnS are widely used infrared (IR) optical materials that are used for IR windows, macro lenses, prisms, and other functional forms. These IR materials are produced by the manufacturers' proprietary chemical vapor deposition (CVD) processing. ZnSe is categorized as a semiconductor material and as a sulfide/chalcogenide. The material appearance is visually characterized by having a yellow–red color and has approximately twice the density of silica. Multispectral ZnS is also a semiconductor material and is a sulfide/chalcogenide with a gray white visual appearance. The material is also approximately 1.5 times the density of silica. Multispectral ZnS has been treated to remove voids and other defects to extend the range of optical transmission to the visible portion of the spectrum. These materials offer a wide band spectral transmission from visible wavelengths through LWIR wavelengths. These materials have an advantage over other IR materials such as silicon or germanium for their ease of optical alignment by using visible band lasers. IR systems that require multispectral imaging or multiple functions such as combining thermal imaging and lidar can be combined into a single optical system using these materials. ZnSe transmits from green light ($\lambda = 0.54\,\mu m$) in the visible spectrum through $17\,\mu m$ wavelength in the LWIR spectral band. Multispectral ZnS transmits from blue light ($\lambda = 0.40\,\mu m$) in the visible spectrum through the middle of the LWIR spectral band ($\lambda = 12.0\,\mu m$). Substrates were purchased from two of the manufacturers of these IR materials, II–VI, Inc., and Rohm and Haas Advanced Materials. The ZnSe and ZnS substrate material from both of these suppliers have been successfully processed with the gray-scale photolithography and reactive plasma etching. There were no observed differences in the plasma etching characteristics of the materials that were sourced from the two suppliers. ZnSe substrates are available in a wide range of sizes and thicknesses, ranging up to 300 mm diameter. Wafers of both materials in 2 and 3 in. diameter form factor have been successfully processed.[23] Cleartran™ from Rohm and Haas is the trade name for the multispectral ZnS.

The optical wafer surface quality for both the ZnSe and ZnS is similar for the materials obtained from both substrate manufacturers. The typical specified substrate surface quality from the manufacturers is 40/20 scratch/dig (S/D). The manufacturers can produce 20/10 S/D quality at a higher wafer cost. Surface roughness or RMS surface roughness is another quality metric for grading the substrates. Wafers are received in and inspected for quality and roughness. Most wafers grade between 20/10 S/D and 40/20 S/D with a few having exceptional surface quality in the better than 10/5 S/D range. Wafer surface roughness is measured at with a ZYGO NewView 5000 white light interferometer. For the ZnSe material, typical surface RMS is from 17 to 25 Å. For Cleartran/Multispectral ZnS, the surface RMS roughness is 17–30 Å. These surface RMS values can be compared to other wafer materials such as prime grade silicon (2–8 Å), 20/10 S/D grade fused silica (5–10 Å), and epitaxial grade GaP (3–10 Å). Figure 2.36 through 2.39 show plots of NewView 5000 scans of the wafer surface prior to processing. The z-height scale depicted in the figures is micrometers (μm).

Note that there is a grainy structure present in all of the surface roughness scans. The appearance of the wafers under a high power microscope using differential interference contrast (DIC) highlights what appears as fracture zones, or grain boundaries. This grainy appearance is visible on nearly all ZnSe or ZnS wafers. One supplier has explained the defined graininess as being subsurface substrate damage that is created during the wafer grinding to thin the wafer to the correct thickness. Polishing is unable to remove all of this subsurface damage. The subsurface damage limits the surface quality or RMS roughness for both of these materials. The roughness does not affect IR applications, but could impact performance on the

FIGURE 2.36 Newview scan of Rohm and Haas ZnSe surface roughness.

FIGURE 2.37 Newview scan of II–VI, Inc. ZnSe surface roughness.

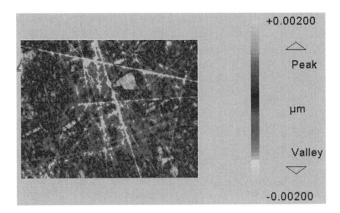

FIGURE 2.38 Newview scan of Rohm and Haas ZnS/Cleartran.

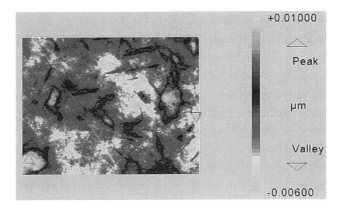

FIGURE 2.39   Newview scan of II–VI, Inc. Multispectral ZnS.

short wavelength high NA optics. This surface artifact will be evident in the plasma etching results.

Plasma etch processes were determined and refined to pattern transfer the 3D photoresist profile into the wafer surface. A DOE investigation was performed to determine which etching parameters were significant to etching the ZnSe wafers. The literature search revealed that most interested parties had experimented with etching ZnS and ZnSe deposited films, rather than bulk substrates.[24–27] The referenced investigators had pursued two significantly different etching chemistries—chlorinated versus methane/hydrogen. From experience with other compound semiconductor material etch development, the methane/hydrogen etching would work, however, processing chamber cleanliness would degrade after only one etch iteration. With the goal being many repeatable deep microlens etches, methane/hydrogen did not appear to be the correct path and was not pursued. During the course of developing plasma etching for other compound semiconductor materials, chlorine processing had yielded significantly higher etching rates. Thus, the chlorinated etch chemistry path was followed. As reported by the authors of Ref. [10], the following reactions are desired to occur in order to etch ZnS and ZnSe:

$$ZnSe + Cl \rightarrow ZnCl_x + SeCl_x$$
$$ZnS + Cl \rightarrow ZnCl_x + SCl_x$$

The DOE traded off reactive gas species such as chlorine with inert gases and passivation gases, processing pressure, RF powers, and wafer temperature. The PlasmaTherm processing tool is used for etching the compound semiconductor materials at MEMS Optical. The tool is an ICP etcher. The system is installed to minimize safety risks from using the corrosive gases and flammable gases. Special gas scrubbers were installed to remove harmful by-products from the processing chamber-pumped exhaust.

Results from the DOE determined substrate etching rates for Cleartran or multispectral ZnS to range from 500 to 3500 Å/min with a selectivity range of 0.35:1 to 2.0:1

where selectivity is defined as the ratio of substrate etch rate to etching mask etch rate. The ZnSe DOE results showed etching rates to range from 770 to 4400 Å/min with selectivity range of 0.35:1 to 3.1:1. For the comparison runs using the identical recipe for identically patterned wafers of both multispectral ZnS and ZnSe, the ZnS exhibited less selectivity and less bulk or substrate etch rate by 20%–30%. The etching exhibits better than 0.5% repeatability in selectivity from run to run. Slight adjustments to various etching parameters permit fine tuning the selectivity in a predictable fashion. The etch process can etch approximately 25 etches (1 wafer cassette) of 20 μm depth per etch before requiring chamber wet cleaning due to build-up in the processing chamber.

2.7.1 Applications for Optical Components in ZnSe and ZnS

Several product applications require the usage of high index of refraction optical materials that are capable of handling higher power loads from IR lasers. Coupling light from a high NA near infrared (NIR) band illumination source into optical fibers is one example application requiring multispectral ZnS. The second example application using ZnSe as a substrate material is for near-field microoptical beam shapers[28] that function at the CO_2 laser lines of 10.6 and 9.6 μm.

The fiber coupling application features a complicated layout with two different lenses simultaneously patterned using gray-scale photolithographic methods. One lens structure is a quad cluster of four tilted aspherical lenses, while the other is an on-axis aspherical lens with a 40% difference in RoC from the tilted aspheric lenses. The back surface of the wafer is also patterned with different lens designs to couple the light into the fiber. Figure 2.40 shows the quad lens and aspherical lens and Figure 2.41 shows an expanded view of the quad lenses. Figures 2.42 and 2.43 show the lenses on the back wafer surface with Figure 2.43 depicting a condensing lens.

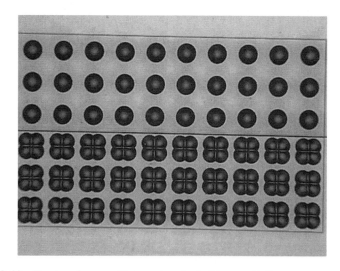

FIGURE 2.40   Front surface quad lens and aspherical lens on ZnS.

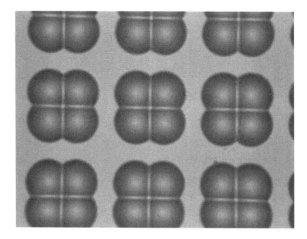

FIGURE 2.41 Expanded view of quad lens on ZnS.

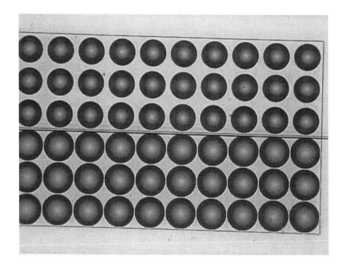

FIGURE 2.42 Back surface aspherical lenses on ZnS.

On the back surface, both lenses were patterned with gray-scale photolithography, and both designs have different aspheric lens design parameters of RoC and conic constant as well as different lens sag heights. Figures 2.40 through 2.43 depict micrographs of aspherical two-sided microlenses manufactured on a Cleartran wafer. The lenses depicted in Figures 2.40 through 2.43 are between 12 and 20 µm in sag height. The lens surfaces and flat area surrounding the lenses are free of pitting and redeposition from the etching process. Some of the subsurface wafer damage prior to processing is evident in the residual surface metrology scans where the lens shape is subtracted from the data and the residual surface is examined for RMS roughness. Figures 2.44 and 2.45 show the wafer subsurface damage present after etching the microlens. The grain structure was present prior to processing as noted above. The

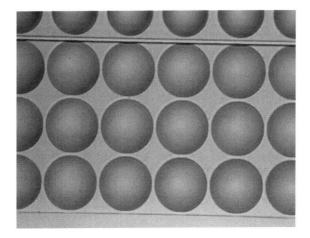

FIGURE 2.43   Expanded view of collimator lens on ZnS.

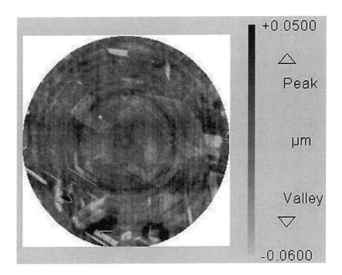

FIGURE 2.44   Surface roughness residual for ZnS collimator lens.

3D lens profiles were measured and the gross lens shape was subtracted to reveal the surface quality of the lens. Figure 2.44 is an image from the collimator aspherical lens on the wafer back surface. The clear aperture represented in the images in Figure 2.44 is 220 μm in diameter. The residual lens surface RMS after subtracting off the lens shape is 200 Å. Some of the noise evident in the residual is from the gray-scale photolithography process. For this particular application, lens surface RMS less than 350 Å provides acceptable insertion losses. Figure 2.45 is the plot of residual roughness for one of the quad aspheric lenses. The clear aperture diameter size is 110 μm and the surface residual RMS is 125 Å.

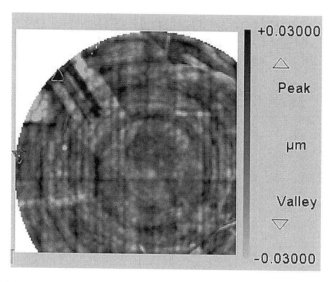

+0.03000

△

Peak

µm

Valley

▽

-0.03000

FIGURE 2.45 Surface roughness residual for ZnS quad lens.

To analyze the uniformity of the manufacturing process, one needs the capability to map the lens design parameters including RoC, conic constant, and surface residual RMS roughness across the etched surface of a processed wafer. The ZYGO MicroLUPI is an instrument that can rapidly measure many lenses in order to characterize the processing uniformity across a wafer. The interferometer-based instrument is a light unequal path interferometer. The collimating lens was measured over the wafer across a 48 by 42 lens grid for 2016 measurements. The approximate mapping time was 4 h. Figure 2.46 shows a graphical depiction of percent deviation from mean RoC as a function of row and column lens position. Figure 2.47 shows the deviation from mean conic value as a function of row and column lens position for ZnS wafer number 03-017_W13FS. The design specification for RoC permits ±5% variation about the mean value. Ninety percent of all RoC values are less than ±2% across the wafer. The design specification for conic constant allows for ±5 conic units. Conic constant is the aspherical term that becomes pronounced off-axis, 5 conic units amounts to 0.5 µm of deviation from the design sag at the edge of the clear aperture. The mapping indicates that all conic values lie within 2 conic units of the mean value.

The second application using ZnSe as a substrate material is for near-field beam shapers operating at the CO_2 laser lines of 10.6 and 9.6 µm. A near-field beam shaper is a smoothly varying diffractive optical element that appears similar to a diffractive lens. The optical design encoded into the diffractive zones remaps the Gaussian input beam to a uniform output. The element can also be used to slightly expand or contract the beam diameter. One optical design solution maps a Gaussian input beam irradiance to a square uniform output irradiance. The second off-axis design maps a Gaussian input beam irradiance to a circular uniform output irradiance. Figure 2.48 shows a micrograph of the etched Gaussian to square near-field beam shaper optic.

03-071 W13FS %RoC Deviation from Mean

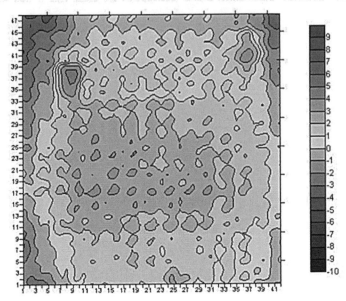

FIGURE 2.46 Graphical representation of wafer mapping RoC for ZnS microlens-filled wafer.

03-071 W13FS Deviation from Conic Mean

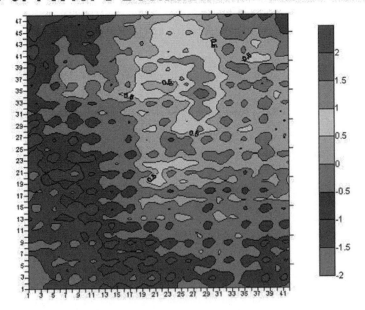

FIGURE 2.47 Graphical representation of wafer mapping of conic constant for ZnS microlens-filled wafer.

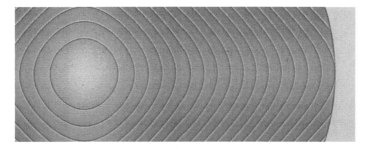

FIGURE 2.48 Micrograph of on-axis Gaussian to uniform square diffractive beam shaping element etched for $\lambda = 10.6\,\mu m$.

FIGURE 2.49 Micrograph of off-axis beam shaper diffractive zones.

The first-order diffractive optic etching depth for a wavelength of 10.6 is 7.5 μm deep. The second design is depicted in Figure 2.49, which is a micrograph showing the zone structure for the off-axis beam shaper element. Figure 2.50 is the same image location as depicted in Figure 2.49 with microscope imaging using DIC filtering enabled to enhance roughness that is present on the blaze zone surfaces. As shown in the surface residual plots for the lens elements, the subsurface substrate damage shows up as having a fractured grainy appearance. Figure 2.51 is a plot of surface residual of one of the zone blazes for the Gaussian to uniform square beam shaper after the blaze shape is subtracted out. The surface residual RMS is 230 Å, which at LWIR wavelengths, has minimal impact on energy transmission through the diffractive optic.

2.8 ETCHING ALTERNATIVE MATERIALS FOR IR APPLICATIONS—IR GLASS (IG6)

An alternative class of materials for IR applications was investigated for applications requiring manufactured microoptical devices[29]. The IR materials considered are referred to as IR glasses. These materials do not have the crystalline or

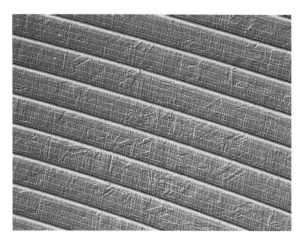

FIGURE 2.50 Micrograph of Figure 2.49 imaged with microcsope DIC filters showing subsurface polishing damage.

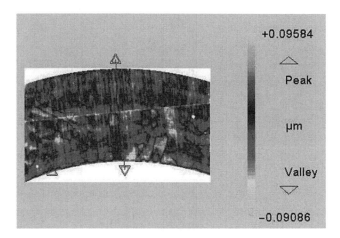

FIGURE 2.51 On-axis Gaussian to uniform square beam shaping element surface RMS residual plot showing subsurface polishing damage.

polycrystalline structure of standard IR optical materials such as silicon, germanium (Ge), GaAs or ZnSe. For these tests an IR glass known as IG6 ($As_{40}Se_{60}$) was considered. Naked Optics Corporation provided the IG6 material in wafer form. The supplied wafers were 50 mm in diameter with a thickness of 2.6 mm. The wafers are received and inspected for quality and roughness. Figure 2.52 shows the measured surface topology for the supplied IG6 wafers. The surface roughness (RMS) is on the order of 25 Å. A cross-sectional slice plot of Figure 2.52 is shown in Figure 2.53. The surface RMS can be compared to the typical surface roughness values of other IR materials such as prime grade silicon (2–8 Å), multispectral ZnS (17–30 Å), ZnSe (17–25 Å), and epitaxial grade GaAs (3–10 Å).

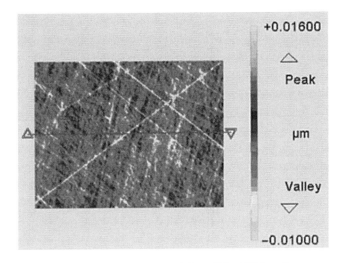

FIGURE 2.52 Measured 180 × 140 µm area of the polished IG6 surface.

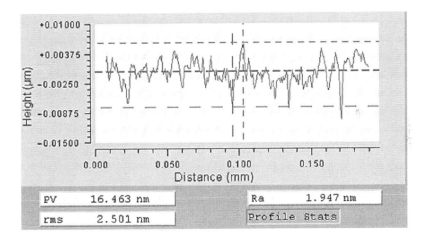

FIGURE 2.53 ZYGO slice profile through the measured area of the polished IG6 surface.

Table 2.4 compares some general physical properties[30] of IG6 to Ge and Si, which are very common optical materials used for the IR spectrum applications. The dispersion characteristics and the low thermal change in refractive index of chalcogenide glasses, such as IG6, allow for the design of color corrected without thermal defocusing when combined with other IR materials in optical systems. Chalcogenide glasses are also available in large wafer diameters allowing for economic advantages to be realized through wafer-scale processing. Typically, the maximum wafer diameter for Ge is 4 and 3 or 4 in. for ZnSe.

TABLE 2.4

Physical Properties of IG6 Compared with Ge and Si

Property	Ge	Si	IG6 ($As_{40}Se_{60}$)
Density (g/cm³)	5.327	2.329	4.63
Thermal expansion ($\times 10^{-6}$ K^{-1})	5.7	2.62	20.7
Specific heat (J/gK)	0.3230	0.7139	0.36
Thermal conductivity (W/mK)	59.9	140	0.24
Transition temperature (°C)	—	—	185
Young's modulus (GPa)	132	162	18.3
Shear modulus (GPa)	54.8	66.2	8.0
Index at 3 μm	4.0445	3.4323	2.8014
Index at 10 μm	4.0044	3.4178	2.7775
dn/dT ($\times 10^{-6}$ C^{-1})	416 at 5.0 μm	159 at 5.0 μm	35 at 3.4 μm
dn/dT ($\times 10^{-6}$ C^{-1})	401 at 20 μm	157 at 10.6 μm	41 at 10.6 μm

2.8.1 Etching Process Development for IG6

Plasma etch processes were determined and refined experimentally at MEMS Optical to pattern transfer the 3D photoresist profile into the wafer surface. For the chalcogenide material, a DOE was run. The development tests consisted of a two-level DOE with midpoints. The main limitation to the number of variables and corresponding test runs was the number of wafer samples available. In this study, three wafers were available, with two being employed for the DOE test, and the third being reserved for testing gray-scale processing of IG6. The PlasmaTherm etcher, shown earlier in Figure 2.2, is used for etching the compound semiconductor materials at MEMS Optical. The ICP separates the tasks of plasma generation from biasing the wafer and allows relatively high etching rates at low wafer and mask damage thresholds. Coarse adjustment of selectivity for chlorine-based material processing is depicted in Figure 2.54. The selectivity of IG6 can be adjusted by changing the ratio of chlorine to boron trichloride processing while maintaining the total mass flow constant. The curve establishes a general trend of increasing selectivity with increasing Cl_2:BCl_3 for a wide range of chamber pressures and bias voltages. The multiple data points along a given gas ratio indicates varied operating settings for other parameters. The coarse curve could be used to set the general selectivity, and a different parameter could define a fine adjustment selectivity curve. The interactions of several variables from the DOE runs can be evaluated in any of several DOE software packages. In this case, the interactions between the processing variables of etching gases, processing pressure, and bias voltage can be analyzed for parameter trends, parameter interactions, and determining a functional model for estimating parameter settings for processing IG6. From this experiment, the functional form contained influences from the three variables being considered.

Results from the DOE determined that etching rates for IG6 range from 13,000 to 64,000 Å/min with a selectivity range of 6.0:1 to 17.0:1. Selectivity greater than 1:1 indicates that the substrate etches more rapidly then the etching mask. Table 2.5 compares the etching properties of IG6 to GaAs, a crystalline IR material of similar etch chemistry. The surface roughness (RMS) after etch pattern transferal when

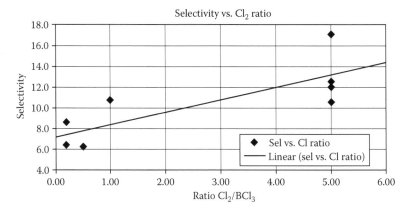

FIGURE 2.54 Selectivity parameter curve for chlorine-based IG6 etching.

TABLE 2.5
Etch Processing Results of IG6 Compared to Results from GaAs

Property	IG6 (As$_{40}$Se$_{60}$)	(GaAs)
Maximum selectivity—lens	17.2:1	5.3:1
Minimum selectivity—lens	6.4:1	0.37:1
Maximum etch rate	6.44 μm/min	3.5 μm/min
Minimum etch rate	1.33 μm/min	0.31 μm/min
Wafer surface (RMS)	Polished 20–30 Å	Epitaxial grade 3–10 Å
Gray-scale Fab etch (RMS)	430–650 Å	30–400 Å
Wafer availability	2 in.–200 mm diameter	2 in.–150 mm diameter
Optical transmission	3.0–12.0 μm	0.95–19.0 μm

using gray-scale photolithography is highly dependent on the initial surface roughness (RMS) of the photoresist. The high selectivity of the etch processing for the IG6 material requires the pre-etch photoresist surface roughness (RMS) to be low. Further process optimization is required to improve the etched surface roughness for MWIR applications. For the LWIR wavelength of 10.6 μm, with a refractive index of 2.778, the one wavelength depth is 5.96 μm. For the 10.6 μm wavelength, the etched surface roughness of 650 Å gives a surface roughness of λ/90.

Figure 2.55 is a picture of a diffractive lens with a diameter of 16.5 mm etched into a 50 mm diameter IG6 substrate using gray-scale manufacturing. Gray-scale photolithography allows for the generation of true kinoform surfaces with high diffraction efficiency. The higher order diffractive etched in Figure 2.55 is 68 μm deep, and was etched at a selectivity of 8:1.

2.9 ETCHING MICROOPTICS IN AN OPTICAL MATERIAL THAT IS NOT REACTIVE

The optical properties of calcium fluoride (CaF$_2$) in the DUV spectral range are particularly interesting for semiconductor photolithography applications. As shown

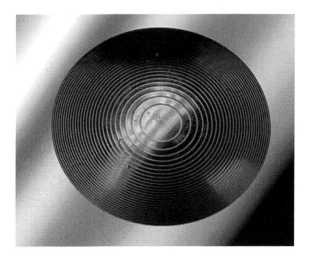

FIGURE 2.55   Zoomed image of gray-scale diffractive lens etched into IG6.

by the spectral transmission curves[31] in Figure 2.56, CaF_2 transmits radiation at a deeper wavelength in the UV regime than does commercially available UV-grade fused silica. CaF_2 is a crystalline material manufactured at very high purity for semiconductor lithography, and has much lower absorption at the DUV wavelengths. This makes CaF_2 an extremely desirable material for microoptical applications in the deep UV, particularly at 157, 193, and 248 nm.

The use of CaF_2 as a microoptic substrate material entails some special considerations. CaF_2 is not reactive to typical reactive ion processing and is therefore not responsive to plasma etching—the method of choice for microoptic fabrication.[32] Hence, the only available etching method is ion milling.[33] Ion milling involves the physical bombardment of the substrate with argon ions, which are accelerated toward the substrate at high speed via the use of an applied electric field,[34] as shown in Figure 2.57. For this application, no reactive gases were used. The ions strike the substrate, dislodging material at the atomic level. This process continues until the substrate has been etched to the desired depth. Surface microroughness from the ion milling process is also a concern, particularly at short wavelengths.[32] Over the course of production development, several countermeasures relating to wafer cooling were implemented to reduce surface microroughness.

2.9.1 GRAY-SCALE FABRICATION OF GAUSSIAN HOMOGENIZER AND MLA

The structures, Gaussian homogenizers and MLAs, were fabricated into the CaF_2 wafers using gray-scale manufacturing.[35] A binary photomask is either opaque or translucent in a photolithography process. A gray-scale photomask allows spatially varying levels of UV transmission through the mask. The gray-scale process flow is shown in Figure 2.58. Features are etched to the proper shape and size through manipulating the resist shape in lithography using the selectivity from the pattern

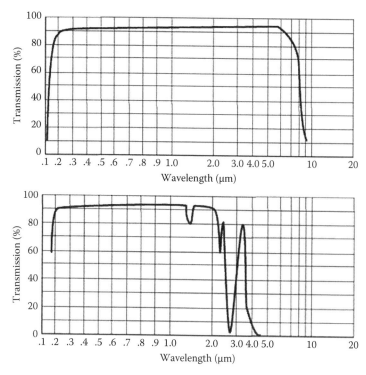

FIGURE 2.56 Transmission curves of CaF_2 (top) and UV-grade fused silica (SiO_2) (bottom).

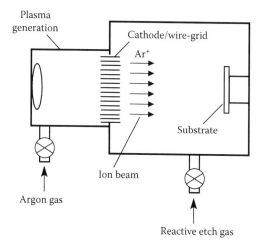

FIGURE 2.57 Schematic of ion mill chamber configuration.

transfer process of the ion mill. In contrast to binary etching, where the primary selectivity concern is to simply retain enough photoresist to protect the nonetched portions of the substrate, gray-scale lithography requires precise control of the lithography feature size and selectivity in order to achieve the desired lens shape.

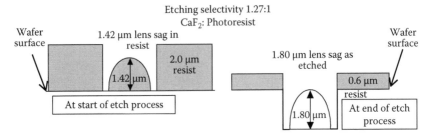

FIGURE 2.58   Gray-scale processing for CaF$_2$ lens etching.

In this presented example, the target sag of the microlens is 1.8 μm. With the etching selectivity process at 1.27:1, the lithography conditions are manipulated to produce a lens having a photoresist sag of 1.42 μm. When etched at 1.27:1 selectivity, the resulting CaF$_2$ lens will have a vertex height of 1.8 μm. Note that only the vertical dimensions are affected by selectivity; the dimensions in x and y about the wafer surface do no change—this must be taken into account with the photomask and lithography process to ensure that when scaled up by the selectivity of the process, the lens will have the correct aspherical shape as defined by the lens parameters of RoC and conic constant. Proper lithography conditions are also essential in producing structures in resist that fall within the etching process window.

The samples included etched gray-scale lenses and homogenizers with depths in the range of 1.0–1.8 μm. Example CaF$_2$ microoptics are shown in Figure 2.59. The CaF$_2$ etch rate was 280 Å/min and the resist etch rate was 220 Å/min for a 1.27:1 selectivity. Etching 1 μm deep features took 70 min of continuous milling. The 1.8 μm deep features were etched for two cycles with adequate cooling between cycles. The use of cooling cycles reduced etching-induced photoresist damage, and reducing microroughness on the lens surface. Typical residual roughness values for the process were 8–20 nm RMS. Sag uniformity within the wafer was ±1.12%, and the wafer-to-wafer sag repeatability was ±1.70%. RoC uniformity within the wafer was ±0.78%, and the wafer-to-wafer RoC repeatability was ±1.89%.

The diffractive optical element manufactured was a Gaussian output homogenizer. The device was designed and etched as a multiorder device to convert a multimode DUV laser input beam to a stable Gaussian output distribution in the far

FIGURE 2.59   Microscope image of CaF$_2$ cylindrical lens array (left), photograph of 10 × 10 × 2.3 mm CaF$_2$ cylindrical lens arrays (right).

FIGURE 2.60 Micrograph of a gray-scale multiorder diffractive Gaussian homogenizer.

field. Figure 2.60 shows the Gaussian output homogenizer. The measured results show the zero-order percentage to be approximately 0.4% and the diffraction efficiency measured at 82%. The output was measured with a waist divergence angle of 3.31°.

ACKNOWLEDGMENTS

The author would like to thank both present and former colleagues for the motivation and inspiration to always extend the limits on what can be accomplished in the fabrication of microoptics. When you believe the limits have been reached on what can be processed, in the not too distant future, these limitations will have to be exceeded. The author would also like to thank the support of past and present customers, colleagues from vendors and equipment manufacturers such as Surface Technology Systems (STS), and colleagues from research institutions such as the staff at Cornell Nanofabrication Facility (CNF). Finally, the author would like to thank his wife Nicole, for her infinite patience and understanding.

REFERENCES

1. C. Wu, High energy beam sensitive glasses, U.S. Patent 5,285,517, February 8, 1994.
2. C. Wu, Gray scale all-glass photomasks, U.S. Patent 6,524,756, February 25, 2003.
3. T. Suleski, W. Delaney, and M. Feldman, Fabricating optical elements using a photoresist formed using a gray level mask, U.S. Patent 6,638,667, October 28, 2003.
4. G. Gal, Method of fabricating microlenses, U.S. Patent 5,310,623, May 10, 1994.
5. W.W. Anderson, J. Marley, G. Gal, and D. Purdy, Fabrication of microoptical devices, *Conference on Binary Optics*, NASA, Huntsville, AL, February 23–25, 1993.

6. D.M. Brown, D.R. Brown, and J.D. Brown, High performance analog profile diffractive elements, *SPIE Proceedings*, Vol. 3633, 46–50, San Jose, CA, January, 1999.

7. L.A. Donohue, J. Hopkins, R. Barnett, A. Newton, and A. Barker, Developments in Si and SiO$_2$ etching for MEMS based optical applications, *Micromachining Technology for Micro-Optics and Nano-Optics II, SPIE Proceedings*, Vol. 5347, 44–53, San Jose, CA, January 27–29, 2004.

8. G.T. Borek, J.A. Shafer, R. Lamarr, and L.R. Simmons, Etching of micro and nano-structures in semiconductor and glass material systems, *Micromachining Technology for Micro-Optics and Nano-Optics III, SPIE Proceedings*, Vol. 5720, 36–47, San Jose, CA, January 25–27, 2007.

9. G. Franz and F. Rinner, Reactive ion etching of GaN and GaAs: Radially uniform processes for rectangular, smooth sidewalls, *Journal of Vacuum Science Technology A* 17(1), 56–61, January/February, 1999.

10. G. Franz, High-rate etching of GaAs using chlorine atmospheres doped with a Lewis acid, *Journal of Vacuum Science Technology A* 16(3), 1542–1546, May/June, 1998.

11. M.E. Ryan, A.C. Camacho, and J.K. Bhardwaj, High etch rate gallium nitride processing using an inductively coupled plasma source, *Physical Status Solidi (A)* 176, 743, 1999.

12. R.J. Shul, G.B. McClellan, S.J. Pearton, C.R. Abernathy, C. Constantine, and C. Barratt, Comparison of dry etch techniques for GaN, *Electronics Letters* 32(15), 1408–1409, July 18, 1996.

13. World wide web link: http://www.ioffe.rssi.ru/SVA/NSM/Semicond/GaN/optic.html; Yu, G., G. Wang, H. Ishikawa, M. Umeno, T. Soga, T. Egawa, J. Watanabe, and T. Jimbo, Optical properties of wurtzite structure GaN on sapphire around fundamental absorption edge (0.78–4.77 eV) by spectroscopic ellipsometry and the optical transmission method, *Applied Physics Letters* 70(24), 3209–3211, 1997.

14. World wide web link: http://www.datanite.com

15. World wide web link: http://www.ioffe.ru/SVA/NSM/Semicond/SiC/optic.html; Shaffer, P.T.B., *Applied Optics* 10, 1034–1036, 1971.

16. M.S. So, S.G. Lim, and T.N. Jackson, Fast, smooth, and anisotropic etching of SiC using SF6/Ar, *Journal of Vacuum Science & Technology B* 28, 2055–2057, September/October, 1999.

17. J.J. Want, E.S. Lambers, S.J. Pearton, M. Ostling, C.M. Zetterling, J.M. Grow, R. Ren, and R.J. Shul, ICP etching of SiC, *Solid State Electronics* 42(12), 2283–2288, 1998.

18. J.D. Scofield, P. Bletzinger, and B.N. Ganguly, Oxygen-free etching of α-SiC using dilute SF$_6$:Ar in an asymmetric parral plate 13.56 MHz discharge, *Applied Physics Letters* 73(1), 76–78, July 6, 1998.

19. J.D. Scofield, B.N. Ganguly, and P. Bletzinger, Investigation of dilute SF6 discharges for application to SiC reactive ion etching, *Journal of Vacuum Science & Technology A* 18(5), 2175–2184, September/October, 2000.

20. C.H. Jeong, D.W. Kim, H.Y. Lee, H.S. Kim, Y.J. Sung, and G.Y. Yeom, Sapphire etching with BCl$_3$/HBr/Ar plasma, *Surface and Coatings Technology* 171, 280–284, 2003.

21. Y.J. Sung, H.S. Kim, Y.H. Lee, J.W. Lee, S.H. Chae, Y.J. Park, and G.Y. Yeom, High rate etching of sapphire wafer using Cl$_2$/BCl$_3$/Ar inductively coupled plasmas, *Materials Science and Engineering B* 82, 50–52, 2001.

22. S.-H. Park, H. Jeon, Y.-J. Sung, and G.Y. Yeom, Refractive sapphire microlenses fabricated by chlorine-based inductively coupled plasma etching, *Applied Optics*, 40(22), 3698–3702, August 1, 2001.

23. G.T. Borek, D.M. Brown, and J.A. Shafer, Gray scale fabrication of microoptics in bulk zinc selenide and bulk multispectral zinc sulfide, *Micromachining Technology for Micro-Optics and Nano-Optics II, SPIE Proceedings*, Vol. 5347, 28–37, San Jose, CA, January 27–29, 2004.

24. J.W. Lee, B. Pathangey, M.R. Davidson, P.H. Holloway, E.S. Lambers, A. Davydov, T.J. Anderson, and S.J. Pearton, Electron cyclotron resonance plasma etching of oxides and SrS and ZnS-based electroluminescent materials for flat panel displays, *Journal of Vacuum Science & Technology A* 16(3), 1944–1948, May/June, 1998.

25. T. Yoshikawa, Y. Sugimoto, Y. Sakata, T. Takeuchi, M. Yamamoto, H. Hotta, S. Kohmoto, and K. Asakawa, Smooth etching of various III/V and II/VI semiconductors by Cl_2 reactive ion beam etching, *Journal of Vacuum Science & Technology B* 14(3), 1764–1772, May/June, 1996.

26. T. Saitoh, T. Yokogawa, and T. Narusawa, Reactive ion beam etching on ZnSe and ZnS epitaxial films using Cl2 electron cyclotron resonance plasma, *Applied Physics Letters* 56(9), 839–841, February 1990.

27. S.H. Su, M. Yokoyama, and Y.K. Su, Characteristics of ZnS thin films etched by reactive ion etching, *Materials Chemistry and Physics* 42, 217–219, 1995.

28. G.T. Borek and D.R. Brown, High performance diffractive optics for beam shaping, *SPIE Proceedings*, Vol. 3633, 51–60, San Jose, CA, January, 1999.

29. J.G. Smith and G.T. Borek, *Etching of Chalcogenide Glass for IR Microoptics, Defense and Security Symposium*, Orlando, FL, 2008.

30. B. Michael, *Handbook of Optics*, Volume 2. McGraw-Hill, New York, 1995.

31. World wide web link: http://www.rmico.com/content/view/103/68/

32. D. Flamm, A. Schindler, T. Harzendorf, and E.B. Kley, Fabrication of microlens arrays in CaF_2 by ion milling, *Proceedings of the SPIE*, 4179, 108–166, 2000.

33. H. Sankur, R. Hall, E. Motamedi, W. Gunning, and W. Tennant, Fabrication of microlens arrays by reactive ion milling, *Miniaturized Systems with Micro-Optics and Micromechanics; Proceedings of the Meeting*, 150–155, San Jose, CA, 1996.

34. S. Sinzinger and J. Jahns, *Microoptics*. Wiley-VCH, Weinhem, 2003, pp. 61–62.

35. J. Lawrence, L. Simmons, A. Stockham, J. Smith, G. Borek, M. Cumme, R. Kleindienst, and P. Weissbrodt, Grayscale homogenizers in calcium fluoride, Advanced Fabrication Technologies for Micro/Nano Optics and Photonics, SPIE Volume 6883, February 7, San Jose, CA, 2008.

3 Analog Lithography with Phase-Grating Masks

Jin Won Sung and Eric G. Johnson

CONTENTS

3.1   INTRODUCTION

During the last decade a revolution has occurred in the optics and photonics industry, largely based on the wafer-level fabrication of components. The key to this process is the lithography and etching of micro-optical elements in a variety of substrates in III–V semiconductors and glass materials. Most processes utilize a patterning and etching process that is repeated a number of times to reach high diffraction efficiencies and is quite costly in wafer-level processing steps. The binary square type of resist profile has been researched and optimized extensively for the past 20 years in order to reach the highest possible level of resolution with the optical stepper of deep ultraviolet (DUV) wavelength. However, the need for single-step lithographic patterning of high-efficiency optical elements is a necessity for high volume applications.

In order to meet this challenge for analog resist profiles, several new types of photomasks have appeared, such as grayscale, halftone, and binary phase masks, for certain applications other than high-resolution integrated circuits (IC) [1–7]. Grayscale masking exploits the continuously varying optical density in the special patented high energy beam sensitive (HEBS) glass plate in order to form a continuous relief profile in the photoresist [2]. Grayscale masks have two main drawbacks. One is high cost, and the other is its strict dependence on the optical density of the photoresist being used. It is necessary to characterize the thickness of the resist in terms of the optical density for a specific exposure tool in order to design a proper optical density map on a grayscale mask. Halftone masks create analog optical transmittance by the use of a square pixel array representing continuous optical density [1,3]. By varying the pixel density or size, halftone masks are capable of creating analog optical transmittance for the incident exposure light. However, this technique suffers from the pixel aperture diffraction effect, and also requires the adjustment of pixel density for a specific exposure tool. Pure binary phase masks are sometimes used for fabricating high-frequency sinusoidal gratings in the photoresist with half the period of the mask. The idea of using the alternating phase shift on the photomask first came up in the early 1980s as a technique of enhancing the resolution by overcoming the diffraction limit of the imaging system [8]. Now, the phase shift mask has become a mature, standard technique for resolution enhancement in the semiconductor industry, and is mainly used for dense periodic patterns of submicron resolution. The phase shift is usually implemented in the alternating opening region of a binary amplitude chrome mask by etching the mask substrate for a phase shift of half wavelength [9].

In this chapter, a new phase mask technique is summarized, which allows for the fabrication of an analog micro-optic profile in both thin and thick photoresists [10–12]. This technique is fundamentally different from the grayscale and halftone mask techniques in that it utilizes a phase function on the mask plane to create analog optical intensity on the wafer plane, while the other two techniques only exploit analog amplitude functions on the mask plane. We investigated the potential of controlling the optical transmittance of the phase mask in a continuous fashion by changing the parameters of the binary phase-grating. Using only the zeroth-order diffraction from the phase-grating with a π phase shift, the optical transmittance can be controlled by simply varying the duty cycle of the phase-grating. We find this technique to be a promising alternative to grayscale and halftone masking techniques in the field of analog photolithography. The feasibility of this technique was investigated and demonstrated in standard photolithographic environment for both one-dimensional (1D) and two-dimensional (2D) optical structures in thick photoresists.

3.2 PHASE MASK BACKGROUND

The optical stepper is a reduction imaging system for a photomask integrated with the condenser system, as illustrated in Figure 3.1. It has a fixed image reduction ratio M and a finite numerical aperture (NA). This finite numerical aperture comes from the pupil aperture inside the stepper system. The radius of the pupil is determined by the numerical aperture of the imaging system. In the case of using a conventional binary amplitude grating mask, there will be several diffraction spots with the zeroth order at the center of the pupil plane and symmetric higher-order spots on the left and right sides, as shown in Figure 3.2. The spacing in spatial frequency f of the adjacent diffraction orders is given by

$$\Delta f = \frac{1}{p} \tag{3.1}$$

where p is the period of grating object on the wafer, and the radius of the pupil is given as NA/λ.

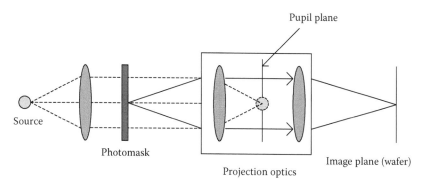

FIGURE 3.1   Optical diagram of the photolithographic stepper. (From Sung, J., et al., *J. Microlith. Microfab. Microsyst.*, 4, 041603-1, 2005. With permission.)

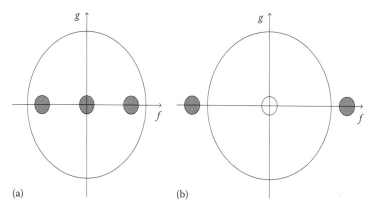

FIGURE 3.2   (a) Pupil diagram of the mask spectrum of binary amplitude grating. (b) Pupil diagram of the binary phase-grating with π phase shift and smaller period. (From Sung, J., et al., *J. Microlith. Microfab. Microsyst.*, 4, 041603-1, 2005. With permission.)

Since the actual stepper system is partially a coherent imaging system, the diffraction spot of each order is a small circular disk of radius σ, indicating the partial coherence factor of the source. It is defined as the ratio of the condenser numerical aperture to the imaging numerical aperture. Its value can vary between 0 and 1, and a smaller value corresponds to a more coherent source [9,11,12]. So, by equating Equation 3.1 with $NA(1 + \sigma)/\lambda$, the cutoff period below, which eliminates all ± 1 and higher orders from the pupil area, is given by

$$p_c = \frac{M\lambda}{NA(1+\sigma)} \tag{3.2}$$

In the above expression, the quantity M is the image reduction factor. This factor is usually 4 or 5, and it means that the wafer image is smaller than the corresponding mask object by this factor. This factor is included to reduce the numerical aperture of stepper on the mask side according to the magnification factor, giving actual cutoff period on the mask plane. Below the cutoff period, the binary amplitude grating would not form an image on the wafer because there is only uniform zeroth-order light intensity. For the grayscale or halftone mask which produces an analog amplitude, the Fourier spectrum on the pupil plane would have some arbitrary analog pattern filling the entire pupil area rather than the discrete diffraction spots, and there is no way to modify this frequency spectrum pattern at the pupil plane. This frequency filtering effect at the pupil is a fundamental property of an optical stepper, which also sets the resolution limit on the feature size [11].

However, if one can make the diffraction efficiency of the zeroth order as a function of position on the mask, then this frequency filtering effect can be utilized to create an analog optical intensity with only a zeroth-order diffracted light from the mask. This can be achieved with a binary phase-grating mask with π phase depth, as shown in Figure 3.3. In this case, the even orders of diffraction spots would not exist if the duty cycle of the grating is 0.5. The duty cycle of a grating is defined as the ratio of line-width to the period. The diffraction efficiency of this type of grating would be a function of duty cycle. The amplitude transmittance of this phase-grating can be expressed in convolution form as

$$t(x) = \left[2\,\text{rect}\left(\frac{x}{a}\right) - 1 \right] \otimes \frac{1}{p}\,\text{comb}\left(\frac{x}{p}\right) \tag{3.3}$$

Taking the analytic Fourier transform of this expression yields the far field diffraction field $T(f)$ as a function of the spatial frequency f:

$$T(f) = [2a\,\text{sinc}(af)/p - \delta(f)]\,\text{comb}(pf) \tag{3.4}$$

By substituting zero in the spatial frequency f and taking the absolute square of the above equation, the zeroth-order diffraction efficiency of this binary phase-grating is given as

$$DE_0(x) = 1 - 4\left(\frac{a(x)}{p}\right) + 4\left(\frac{a(x)}{p}\right)^2 \tag{3.5}$$

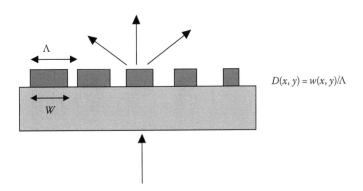

$$D(x, y) = w(x, y)/\Lambda$$

FIGURE 3.3 The binary phase-grating mask with spatially varying duty cycle.

where p is the period and a is the width of grating line which can be varied as a function of position on the mask plane.

At the half-duty cycle a becomes $p/2$, and the diffraction efficiency becomes zero. But, as w departs from the half-duty cycle, the value of DE_0 begins to rise in the form of a parabolic function as shown in Figure 3.4.

By varying the line-width a as a function of position on the mask, the exposure intensity passing through the mask would also be a function of position on the mask. Therefore, it is possible to create any arbitrary analog exposure intensity by designing the appropriate duty cycle function, $W(x) = a(x)/p$, for the binary phase-grating mask. Equation 3.5 of the zeroth-order diffraction efficiency can be considered as the desired target intensity transmittance, $I(x)$, coming out of the mask. The duty cycle, $W(x) = a(x)/p$, can then be stated as

$$W(x) = \frac{1}{2}\left(1 - \sqrt{I(x)}\right) \tag{3.6}$$

By knowing the desired analog intensity transmittance, $I(x)$, for a certain micro-optical element, it is possible to design a phase-grating with the corresponding duty cycle.

The GCA g-line stepper used in this research has a source wavelength of 436 nm, 0.35 of numerical aperture and 0.6 of partial coherence factor, giving 0.78 μm of minimum resolvable period on the wafer. The actual cutoff period of the binary grating on the mask for a 5× stepper is five times bigger to be 3.89 μm. So, if the period of the binary grating on the mask is smaller than the cutoff period, the ±1

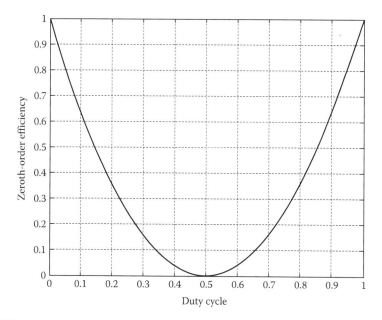

FIGURE 3.4 Zeroth-order diffraction efficiency of binary phase-grating of π phase depth as a function of duty cycle. (From Sung, J., et al., *J. Microlith. Microfab. Microsyst.*, 4, 041603-1, 2005. With permission.)

and the higher-order diffracted light from the mask would be cut off by the pupil aperture at the stepper, and only the zeroth order will contribute to form uniform flat intensity on the wafer. This is the resolution limitation for a binary amplitude grating mask.

In order to create analog resist profiles with a photomask, the intensity profile coming through the mask should be made continuous. This can be achieved by creating an analog light amplitude with a halftone mask [3] as illustrated in Figure 3.5. This type of mask uses a set of subresolution opaque pixels and a fixed center to center subresolution period (pitch). Since the period is subresolution, smaller than the value obtained from Equation 3.2, only the zeroth-order diffraction will get through the stepper system to form the image on the wafer. The imaging intensity on the wafer plane is proportional to the relative amount of the area not blocked by the opaque pixels and can be calculated with the following formula:

$$I = 1 - \left(\frac{A_{pixel}}{A_{pitch}} \right) \tag{3.7}$$

In this equation, A_{pixel} and A_{pitch} represent the area of pixel and pitch, respectively.

Thus, by varying the area of pixels as a function of position on the mask, it is possible to create analog intensity profiles coming out of the mask and fabricate a desired analog resist profile with the proper resist exposure and development time.

A drawback of this approach for an analog resist profile is that it is plagued by the light scattering occurring at the edge of pixels. This is due to the fact that there is a sharp transition in light amplitude from 0 to 1 at the boundary of the chrome pixel. This scattered light can be stray light in the optical stepper system and contribute

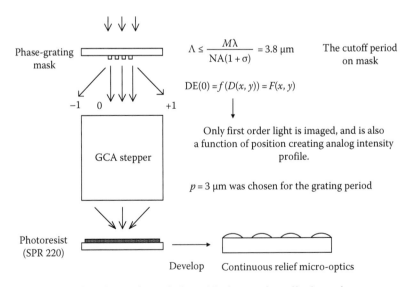

FIGURE 3.5   Analog photoresist sculpting with phase and amplitude gratings.

to degradation in the smoothness of the imaging intensity. In order to avoid this stray light problem, the analog intensity profile should be created without using the binary chrome pattern like the halftone mask. To address this issue we came up with the idea of using a phase only binary grating mask with π phase depth [10]. The layout of this mask is similar to the halftone mask as shown in Figure 3.6, but each pixel square has a π phase shift relative to the background area instead of opaque chrome. The phase shifting pixels can be made by patterning a photoresist which is transparent for the stepper wavelength. If the period of this 2D phase-grating is smaller than the value of 2 and the phase shift of the pixels is π, then only the zeroth order will go through the stepper system. Furthermore, the intensity of the zeroth-order diffracted light will depend on the fill factor of the pixel. So, the size of this square pixel can be varied to control the zeroth-order efficiency from this grating. The fill factor is defined as the ratio of area of pixel to area of pitch, and is also related to the duty cycle as the following expression:

$$F = w^2 = \frac{a^2}{\Lambda^2} \tag{3.8}$$

The amplitude of this 2D binary phase-grating with π phase shift can be expressed in the following form:

$$t(x,y) = \left[2\,\mathrm{rect}\left(\frac{x-a/2}{a}, \frac{y-a/2}{a} \right) - 1 \right] \otimes \frac{1}{\Lambda^2}\mathrm{comb}\left(\frac{x}{\Lambda}, \frac{y}{\Lambda} \right) \tag{3.9}$$

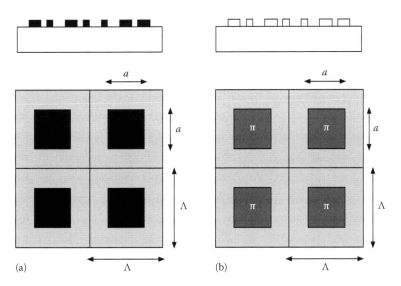

(a)   (b)

FIGURE 3.6 Two different types of photomasks for analog resist profile formation. (a) Halftone mask. (b) Binary phase-grating mask. (From Sung, J., et al., *Appl. Opt.*, 45, 33, 2006. With permission.)

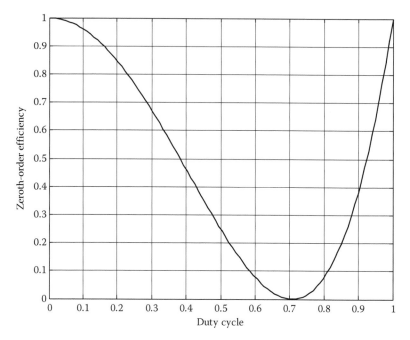

FIGURE 3.7 Zeroth-order efficiency vs. duty cycle curve for 2D binary phase-grating. (From Sung, J., et al., *Appl. Opt.*, 45, 33, 2006. With permission.)

By taking the 2D Fourier transform of this expression and putting (0, 0) in the spatial frequency, we arrive at the zeroth-order efficiency equation below:

$$DE(w) = 1 - 4w^2 + 4w^4 \qquad (3.10)$$

The zeroth-order efficiency vs. duty cycle plot for Equation 3.10 is nonsymmetric with zero at $w = 1/\sqrt{2} \cong 0.707$ as shown in Figure 3.7. This is due to the fact that the 2D binary phase-grating makes the zeroth-order efficiency zero at the half-fill factor ($F = 1/2$). The cutoff period for the 2D binary phase-grating is given by Equation 3.2, but with the factor of $1/\sqrt{2}$ due to the presence of diagonal diffraction orders.

This 2D efficiency equation can be solved for the duty cycle as well. So the 2D grating duty cycle $w(x, y)$ as a function of position (x, y) on the mask can be obtained from the desired zeroth-order intensity transmittance profile $I(x, y)$. Thus, it is possible to design the 2D binary phase-grating mask for arbitrary 2D analog resist profiles.

3.3   DESIGN AND FABRICATION OF OPTICAL ELEMENTS

3.3.1   Photoresist Characteristics

In order to properly design the duty cycle of a binary phase-grating mask for the desired analog resist profile, the exposure response of the thick photoresist should be characterized in terms of the duty cycle of the binary phase-grating mask. The SPR

220-7 photoresist can be spun to form 7 to 12 µm of thickness on a fused silica wafer with good uniformity; it also works well for both I (365 nm) and g-line (436 nm) exposure tools. It was spun on a fused silica wafer at the thickness of 12 µm and exposed with a GCA g-line stepper for exposure times of 0.3–3.6 s in a dose matrix form on the wafer, as illustrated in Figure 3.8. Since the exposure intensity at the wafer in our stepper is 150 mW/cm², it corresponds to an exposure dose range of 45–540 mJ/cm². This wafer was developed in an MF CD-26 developer for 4 min, and the developed depth (µm) vs. exposure dose (mJ/cm²) was plotted, as in Figure 3.9. The SPR 220-7 resist responds slowly to the delivered exposure energy until the

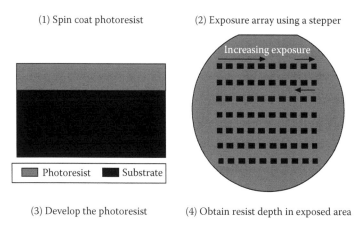

(1) Spin coat photoresist (2) Exposure array using a stepper

(3) Develop the photoresist (4) Obtain resist depth in exposed area

FIGURE 3.8 Dose exposure characterization process.

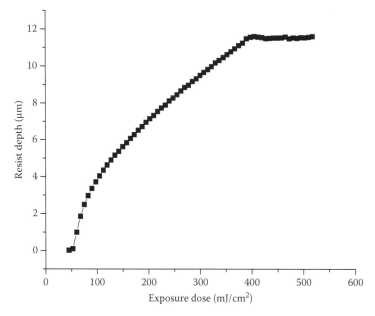

FIGURE 3.9 Developed thickness of SPR 220-7 resist vs. exposure time in stepper. (From Sung, J., et al., *Appl. Opt.*, 45, 33, 2006. With permission.)

exposure time reaches 0.5 s (75 mJ/cm²) and begins to rise rapidly past that point. After 0.8 s of exposure time, the slope of the exposure curve is slightly reduced and maintains a linear form until it reaches saturation at around 2.8 s (420 mJ/cm² of dose). Because of this nonlinear response of the resist to the exposure dose, a proper amount (0.6–0.8 s) of flat bias exposure is necessary prior to the exposure with the phase mask. After this bias exposure without the mask, the SPR 220-7 responds to the delivered analog dose in a smooth, almost linear fashion.

It is possible to convolute numerically the zeroth-order efficiency vs. the duty cycle curve of the phase-grating with the exposure curve, in order to make a curve of remaining resist thickness d (μm) vs. duty cycle resulting from the sequential bias exposure and the exposure through the phase mask. This convolution curve is very useful for estimating the response of the resist to the duty cycle of the phase-grating mask. It also helps to determine the appropriate range of duty cycles for a linear response in the resist profile.

In order to perform this numerical convolution, the scales of the efficiency curve and exposure curve should be matched. The zeroth-order efficiency $DE(w)$ as a function of the grating duty cycle w can be regarded as the intensity transmittance of exposure light, which is the ratio of the zeroth-order intensity $I(w)$ to the flat exposure intensity I_0. In order to obtain the absorbed dose, it should be multiplied by the flat exposure intensity I_0 and exposure time t (s). Also, there should be an additional term for the flat bias exposure, which is done prior to the exposure with the phase-grating mask. This is simply the product of the flat exposure intensity I_0 (150 mW/cm²) and the bias exposure time, t_b (s). Thus, the total amount of absorbed dose $D(w)$ (mJ/cm²) as a function of duty cycle w is given by

$$D(w) = I_0(t_b + DE(w)t) \qquad (3.11)$$

The next step is the polynomial curve fitting of the exposure characteristic curve. The developed depth h (μm) vs. exposure dose D (mJ/cm²) can be fitted to a sixth-order polynomial $h(D)$ of D. Once we have the coefficients for this sixth-order polynomial, we can simply substitute the dose parameter with $D(w)$ of Equation 3.11, and then it becomes the curve $h(w)$ of developed depth h (μm) vs. the duty cycle w. The remaining resist curve $d(w)$ (μm) is simply the initial thickness d_0 minus the developed depth curve $h(w)$. Figure 3.10 is the remaining resist thickness $d(w)$ resulting from this convolution for 0.6 s of bias and 2.6 s of exposure with a 2D binary phase-grating mask. As shown in this curve, the 2D square phase-grating mask makes a rapid variation of the remaining thickness with the duty cycle of the phase-grating mask. This property makes the 2D phase-grating mask ideal for high sag micro-optics.

3.3.2 PHASE MASK DESIGN

In order to demonstrate the feasibility of this binary phase-grating-mask approach for the fabrication of analog micro-optics, a simple square microprism of 100 μm width was designed. The height of the microprism was chosen to be 7 μm and it had 0.5 μm of base thickness. From an optical viewpoint, this base thickness is redundant and is unnecessary, but if the resist height goes to zero at the edge of the element, the

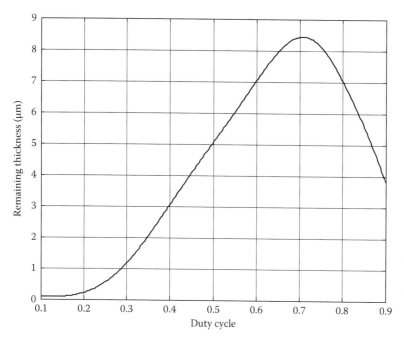

FIGURE 3.10 Remaining thickness vs. duty cycle obtained by numerical convolution of the resist characterization curve with zeroth-order efficiency curve of binary 2D phase-grating for 0.6 s of bias and 2.6 s of exposure time. (From Sung, J., et al., *Appl. Opt.*, 45, 33, 2006. With permission.)

corresponding grating duty cycle on the mask would be almost zero, which is practically impossible. We chose 0.5 μm as the minimum line-width for the phase-grating mask, which corresponds to 0.24 of duty cycle and 0.6 μm of actual pixel size for a 2D phase-grating mask with 2.5 μm of period. According to the convoluted remaining resist curve, this would result in 0.5 μm of remaining resist, giving the microprism the desired resist pedestal. Following the same procedure, we also designed a V-groove element with a 7 μm depth and a total width of 100 μm.

There are two approaches to the design of the duty cycle map of the binary phase-grating for the fabrication of a desired resist profile. The first is to just start from the zeroth-order grating transmittance profile that is of the reverse form as the desired resist profile. This is the simplest way of designing the duty cycle and it assumes that the response of the resist to the exposure intensity is linear. For the microprism profile like this example, the transmittance profile should be of inverted linear ramp form with the width. From the transmittance profile $I(x, y)$, the duty cycle map $W(x, y)$ can be computed by solving the grating efficiency Equation 3.10 for W:

$$W(x, y) = \sqrt{\frac{1 - \sqrt{I(x, y)}}{2}} \tag{3.12}$$

The above expression for the duty cycle function is for lower duty cycle solutions of the grating efficiency equation. For higher duty cycle solutions, the minus sign in the equation should be changed to plus. When designing the desired grating transmittance profile, it should be normalized properly to avoid the minimum duty cycle becoming too small to fabricate. It was normalized to 0.78 of the maximum transmittance, which corresponds to 0.24 of the minimum duty cycle for the 2D grating mask.

The second approach is to use the numerical deconvolution method to find out the required dose profile $D(x, y)$ (mJ/cm^2) for the desired 2D resist height profile $d(x, y)$ (μm) with a certain bias and exposure time, and compute the intensity transmittance $I(x, y)$ and duty cycle profile $W(x, y)$ from that dose profile. The desired 2D resist height profile $d(x, y)$ (μm) can be used to determine the required exposure dose profile to produce it using the polynomial fitting expression for the exposure curve $h(D)$ (μm) of SPR 220 resist. The exposure characteristic curve in Figure 3.3 can be put on the exposure dose D (mJ/cm^2) vs. developed resist depth h (μm) scale and represented by the sixth-order polynomial function $D(h)$:

$$D(h) = a_0 + a_1 h + a_2 h^2 + a_3 h^3 + a_4 h^4 + a_5 h^5 + a_6 h^6 \qquad (3.13)$$

This function gives the required dose value D (mJ/cm^2) for a certain developed resist depth h (μm). Since the relation between the developed resist depth h and the desired remaining resist height d is $d_0 = h(x, y) + d(x, y)$ for the total resist thickness d_0, it is possible to substitute the desired resist depth profile $h(x, y) = d_0 - d(x, y)$ in the polynomial expression $D(h)$ in order to obtain the desired dose profile $D(x, y) = D(h(x, y))$. Equation 3.11 relating the exposure dose and intensity is still valid with the spatial coordinate parameters (x, y) in place of the duty cycle w. Thus, with a proper bias and exposure time to clear the SPR 220 resist, the desired dose profile can be used to compute the required transmittance profile by solving Equation 3.11 for $I(x, y)$:

$$I(x, y) = \frac{(D(x, y)/I_0 - t_b)}{t_e} \qquad (3.14)$$

From the exposure curve of SPR 220-7, we determined the bias and exposure time to be 0.6 and 2.6 s, respectively. Finally, the duty cycle distribution $W(x, y)$ can be obtained from the required transmittance profile $I(x, y)$ using Equation 3.12.

In order to verify this principle and the design method for the phase-grating mask for analog resist elements, we first designed simple prism profiles, which are basically linear ramp in the resist profile. The size of the prism is 100 μm square, and height is 7.5 μm. Both transmittance and resist profile-based 2D phase-grating mask design was used to fabricate this structure in SPR 220-7 resist. The target transmittance profile is a linear ramp of transmittance going from 1.66% to 78%. For the resist profile-based phase mask design, the linearly ramping resist profile going from 0.5 to 7.5 μm of resist height was used. Next, we designed a V-groove structure using 2D binary phase-grating mask designed from both the transmittance and resist profiles. For transmittance profile-based phase mask design, an inverted V-type profile with 0.78 of transmittance at the central top was used to generate a duty cycle map with Equation 3.12. The resist profile-based phase mask design was obtained from the V-like trough profile with 7 μm of depth and 100 μm of width using the numerical deconvolution procedure for extracting the duty cycle map explained previously. The ranges of intensity variation and duty cycle are similar to the prism cases. The 1D transmittance profiles and duty cycle maps for the phase-grating mask designed from the resist profiles of the microprism and the V-groove are shown in Figure 3.11.

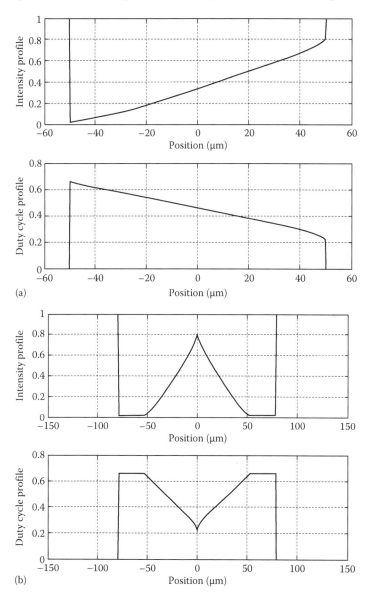

FIGURE 3.11 The intensity transmittance and duty cycle profiles of phase-grating mask (a) for 100 μm microprism and (b) for V-groove. (From Sung, J., et al., *Appl. Opt.*, 45, 33, 2006. With permission.)

Once we have designed the duty cycle map $W(x, y)$ for the 2D phase-grating mask using the resist profile-based approach, it is possible to predict the developed analog resist profile analytically by utilizing the similar numerical convolution method used for producing the curve of Figure 3.10. We can simply put the duty cycle function $W(x, y)$ into Equation 3.10 to obtain the intensity transmittance profile $I(x, y)$ coming out of this mask. Then, we can use Equation 3.14 to obtain the exposure dose profile $D(x, y)$ (mJ/cm^2) from the intensity profile $I(x, y)$ with a certain bias time t_b and exposure time t. It is then straightforward to perform the numerical convolution with the exposure curve of the developed resist depth h (µm) vs. dose D (mJ/cm^2) to finally obtain the developed resist profile $d(x, y)$ (µm). In the case of the phase-grating mask designed from the transmittance profile, this procedure of predicting the developed resist profile analytically is much simpler as the intensity transmittance profile $I(x, y)$ does not need to be computed. The analytically predicted microprism and V-groove profiles were compared with the target profile as shown in Figure 3.12. As is evident from these figures, the transmittance-based phase-grating mask produces a larger mismatch on the lower side edge of the element than the phase-grating mask based on the resist profile. This is due to the fact that the zeroth-order intensity transmittance of Equation 3.10 becomes rapidly nonlinear near the minimum duty cycle region, where the intensity is high and remaining resist is low. However, the predicted resist profile of the resist profile-based phase-grating mask makes excellent agreement with the target resist profile.

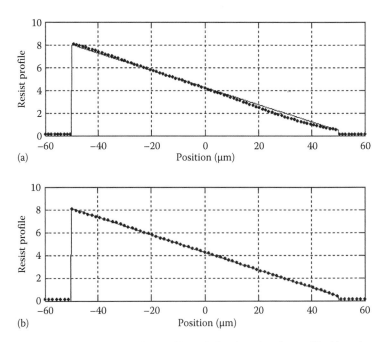

FIGURE 3.12 The 1D target resist profile and developed resist profile (dotted curve) of microprism resulting from the numerical convolution of the exposure curve of SPR 220 and designed duty cycle map. (a) From intensity transmittance-based phase-grating mask. (b) From resist profile-based phase-grating mask. (From Sung, J., et al., *Appl. Opt.*, 45, 33, 2006. With permission.)

3.3.2.1 Phase Mask Fabrication

The phase-grating masks for the 100 μm microprism and V-groove elements were designed with the 2D square pixel phase-grating, as was discussed. The grating period p was chosen to be 2.5 μm according to the cutoff Equation 3.2. First, the desired duty cycle map was created in a MATLAB® code according to the above numerical procedure and the resulting numerical matrix of the duty cycle was saved as an ASCII file. We developed a GDS2 file writing program with the C language, and ran this program with the ASCII text file to create the GDS2 mask file of binary phase-grating mask. It is possible to fabricate the phase-grating mask with a conventional photolithographic technique using a chrome mask and UV exposure tools like the stepper and the mask aligner. However, e-beam direct writing with the poly(methyl methacrylate) (PMMA) resist has proven to be the best way in terms of low absorption and cleanness of the fabricated grating surface. We used 495 A6 from MicroChem for the e-beam resist and coated a 5 in. fused silica mask plate with it for a thickness of 420 nm. This thickness makes a π phase shift at the g-line (436 nm). The e-beam writing was performed with a Leica EBPG5000+ Electron Beam Lithography System. The optimum e-beam dose for clearing the PMMA resist was found to be 450 μC/cm². After the e-beam writing was finished, the mask plate was developed in the methyl isobutyl ketone:iso-propyl alcohol (MIBK:IPA) (1:3) developer for 70 s and rinsed with IPA. A microscopic picture of the fabricated phase-grating mask for a microprism is shown in Figure 3.13.

3.3.3 MICRO-OPTICS PHOTORESIST PROCESSING

For the fabrication of these analog elements on the SPR 220-7 resist with our fabricated phase-grating mask, we coated a 4 in. fused silica wafer with SPR 220-7 resist for an initial thickness of 12 μm. This is the maximum thickness possible with the SPR 220-7 resist when spun at 1000 rpm with a spin coater. Next, it was soft-baked on a hot plate at 115°C for 90 s. We used a GCA g-line stepper to expose this wafer with

FIGURE 3.13 A microscope image of the phase-grating mask fabricated on PMMA e-beam resist using e-beam direct writing technique. (From Sung, J., et al., *Appl. Opt.*, 45, 33, 2006. With permission.)

the phase-grating mask. The bias and exposure times were 0.6 and 2.6 s, respectively, as the duty cycle map on this mask was designed for that exposure condition. The exposed SPR 220-7 resist should sit for at least 45 min before the postbake is applied, since it takes this length of time to finish and stabilize the photochemical reaction. The postexposure bake was done in the same way as the soft-bake. After the postexposure bake was done and the wafer cooled down, it was developed by being immersed in the MF CD-26 developer for 4 min. Last, it was rinsed with double-ionized (DI) water and dried with nitrogen.

The result of the profile measurement shows good agreement with the predicted resist profile obtained from numerical convolution method. We compared the 1D numerically predicted resist profile with the fabricated resist profiles of the prism and the V-groove as shown in Figure 3.14. Both elements were obtained using the resist profile-based phase-grating mask. The slightly higher experimental profile on the low side of the prism is due to somewhat oversized small phase-grating pixels in those areas. Using the numerical convolution computation with the introduction of the pixel size error, this oversizing error of the grating pixels was estimated to be 3%–4%. This oversizing is likely caused by the slight overdeveloping of the PMMA resist after e-beam writing. Both the transmittance-based phase-grating design and the resist profile-based phase-grating design produce final resist profiles in good agreement with the predicted resist profile. However, in the case of

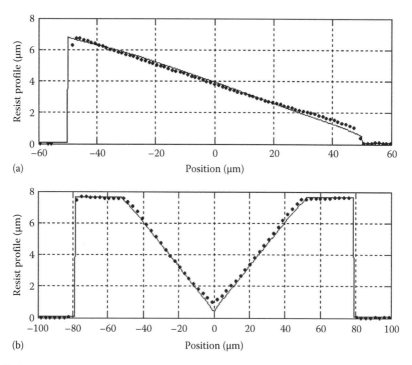

FIGURE 3.14 The comparison of the numerically predicted 1D microprism profile and the fabricated 1D microprism profile (dotted curve): (a) microprism and (b) V-groove. (From Sung, J., et al., *Appl. Opt.*, 45, 33, 2006. With permission.)

the transmittance-based phase-grating mask, the profile near the bottom edge of the prism creates a larger mismatch to the designed prism profile than the profile from the resist-based phase-grating mask. Because the linear ramp in the intensity profile does not produce the linear resist profile as a result of the nonlinear exposure response of the SPR 220 resist, there is more deviation from the designed prism profile than the prism profile made using the resist profile-based phase-grating mask. Thus, it is evident that the resist-based phase-grating mask should be used in order to fabricate the exact desired analog resist element.

We also designed and fabricated analog vortex elements with resist profile-based phase-grating masks. The vortex is the analog surface profile ramping up in angular fashion. The analytic expression for the vortex of height d_0 is given by

$$d(x, y) = d_0 \frac{\tan^{-1}(my/x) + \pi}{2\pi} \tag{3.15}$$

In the above expression, m is the integer charge number for vortex. For $m = 1$, the vortex profile makes one circular ramping from $0°$ to $360°$. For $m = 2$ or higher integers, there will be two or more angular ramping profiles in $360°$. We used Equation 3.15 with $7\,\mu m$ of the total height value d_0, and charge number m equal to 1 and 3 in order to design and fabricate the resist-based phase-grating mask, as explained previously. Microscope images of the phase mask for both $m = 1$ and $m = 3$ charge numbers are illustrated in Figure 3.15a and b, respectively. The resulting analog vortex profiles after exposure and development were shown in Figure 3.16. We checked the angular linearity of the vortex profile by sampling the points lying on a circular and it was in fact a linear relationship with the angle as one would expect.

3.4 DESIGN AND FABRICATION OF AXIALLY SYMMETRIC ELEMENTS

In case of radially symmetric elements like microlenses, it is possible to use the radius coordinate $r = \sqrt{x^2 + y^2}$ as the 1D coordinate thanks to rotational symmetry. The phase-grating mask for the rotationally symmetric element would be a circular ring type of grating, and the corresponding phase profile function can also be represented as a function or r. Thus, all previous expressions can be used with the coordinate r in place of x. Although this is actually a 2D feature, it can be represented with one variable and can be fabricated using a 1D rotationally symmetric phase-grating mask. As an example, consider a lens that has rotational symmetry; the corresponding intensity exposure follows the opposite of the desired surface profile in intensity with an array of concentric rings of variable duty cycles, as illustrated in Figure 3.17. Using the experimental exposure curve of SPR 220 resist and numerical convolution of the phase-grating efficiency with the resist characteristic curve, a circular phase-grating mask was designed for the fabrication of positive spherical microlenses and ring lenses. We also designed the phase-grating mask using a simple intensity profile-based method. For the intensity profile-based mask design, an inverted spherical intensity profile was used as the target intensity. The diameters

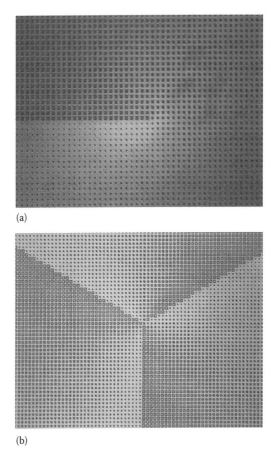

(a)

(b)

FIGURE 3.15   Phase masks for vortex micro-optics: (a) $m = 1$ charge and (b) $m = 3$ charge.

of positive lens and ring lens are 100 and 200 μm, respectively. The shape of a ring lens is like a donut and its 1D profile from the center to the edge is the same as the positive lens. The target sag of the microlens was chosen to be 5 μm by considering the maximum variation of thickness practically possible from the resist convolution curve of Figure 3.18. This corresponds to the duty cycle variation from 0.15 to 0.45 and the grating line-width variation from 0.45 to 1.35 μm on the mask scale. The period of the final phase mask pattern was designed to be 3.0 μm, less than the cutoff period for a GCA stepper (3.7 μm).

The designed binary phase-grating mask was fabricated using the e-beam direct writing technique. PMMA e-beam resist was coated at the π phase thickness for $\lambda = 436$ nm. The mask was then written on a Leica EBPG5000+ Electron Beam Lithography System and developed with an MIBK:IPA (1:3) for 70 s using the immersion technique (see Figure 3.19). The mask was then immediately placed in the stepper, ready for use in creating analog micro-optics. Other mask fabrication methods have been utilized for the creation of the phase-grating mask, including contact and

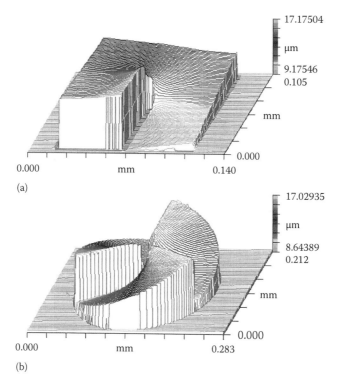

FIGURE 3.16 Two-dimensional Zygo profiles of fabricated analog vortex elements on SPR 220 resist: (a) $m = 1$ vortex and (b) $m = 3$. (From Sung, J., et al., *Appl. Opt.*, 45, 33, 2006. With permission.)

projection lithography. However, the e-beam direct writing technique has shown the greatest consistency, accuracy, and smoothness in the phase-grating formation.

The fabricated binary phase-grating mask was used to expose fused silica wafers coated with 12 µm thick SPR 220 photoresist. In order to fabricate the positive microlens and ring lens on the SPR 220 resist with the fabricated phase-grating mask, we coated a 4 in. fused silica wafer with SPR 220 resist for the initial thickness of 12 µm. Next, it was soft-baked on a hot plate at 115°C for 90 s. A GCA g-line stepper was used to expose this wafer with the phase-grating mask. The bias and exposure times were 0.7 and 2.7 s, respectively, as the duty cycle map on this mask was designed for that exposure condition. The exposed SPR 220 resist then sat for 45 min per the resist requirements before the postexposure bake was applied. The postexposure bake was done in the same way as the soft-bake. After the postexposure bake the wafer was allowed to cool down; it was then developed by being immersed in the MF CD-26 developer for 4 min. Last it was rinsed with DI water and dried with nitrogen.

The surfaces of the fabricated microlenses and ring lenses were measured with a Zygo white light interferometer, and the 2D lens surface figures from both the intensity and resist profile-based designs of the mask are shown in Figure 3.20. The

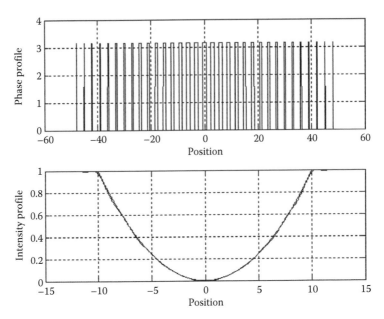

FIGURE 3.17 A simple circular binary phase-grating profile (top) for 20 μm microlens and the computed aerial image (bottom). (From Sung, J., et al., *J. Microlith. Microfab. Microsyst.*, 4, 041603-1, 2005. With permission.)

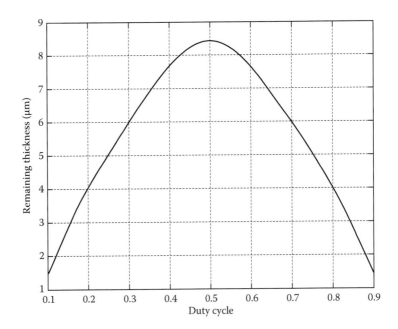

FIGURE 3.18 Remaining thickness of SPR 220 photoresist obtained by convolution process. (From Sung, J., et al., *J. Microlith. Microfab. Microsyst.*, 4, 041603-1, 2005. With permission.)

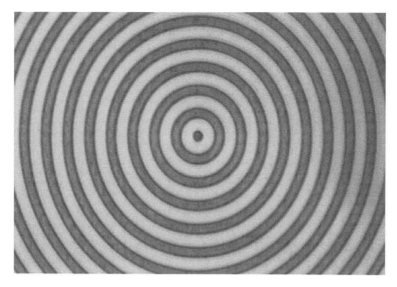

FIGURE 3.19 Fabricated phase mask for positive microlens. (From Sung, J., et al., *J. Microlith. Microfab. Microsyst.*, 4, 041603-1, 2005. With permission.)

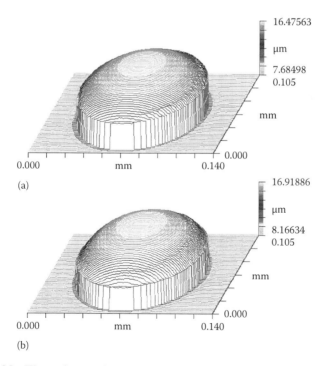

FIGURE 3.20 The surface profiles of fabricated microlenses from Zygo white light interferometer. (a) Microlens fabricated from intensity-based phase-grating mask. (b) Microlens fabricated from resist profile-based phase-grating mask. (From Sung, J., et al., *J. Microlith. Microfab. Microsyst.*, 4, 041603-1, 2005. With permission.)

total height and sag of the lens were 8 and 5 μm, respectively, matching the design. However, it is apparent that the microlens made with the resist profile-based phase-grating mask design has a more accurate spherical surface profile than the other microlens made with the intensity-based phase-grating mask design. To compare the fabricated lens profiles with the designed spherical lens surface, we fitted the lens surface to the following 2D aspheric equation up to the fourth order as a function of the radial coordinate r:

$$d(r) = \frac{r^2/R}{1 + \sqrt{1 - (1+k)r^2/R^2}} + a_4 r^4 \qquad (3.16)$$

From the above equation, the aspheric lens surface parameters such as the radius of curvature R, the conic constant k, and the fourth-order aspheric coefficient a_4 were obtained. Using these lens parameters obtained from the surface fitting, we plotted the experimental aspheric lens profile and compared it with the designed spherical surface. Figure 3.21 shows these plots for both lenses made with two different designs of the phase-grating mask. The microlens fabricated with the resist

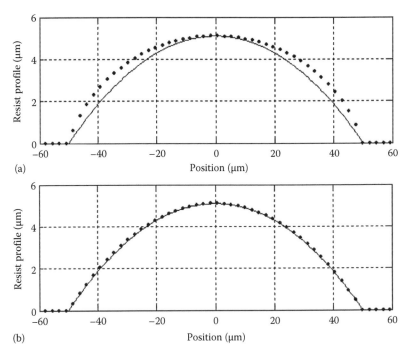

FIGURE 3.21 Targeted lens profile (line curve) and reconstructed 1D lens profile obtained from lens surface fitting to aspheric equation (dotted curve). (a) Microlens fabricated from intensity-based phase-grating mask. (b) Microlens fabricated from resist profile-based phase-grating mask. (From Sung, J., et al., *J. Microlith. Microfab. Microsyst.*, 4, 041603-1, 2005. With permission.)

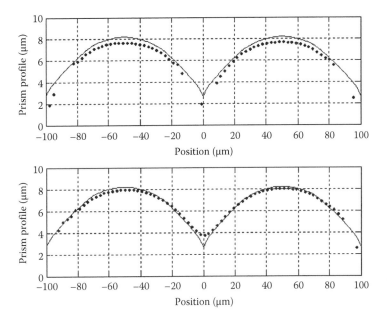

FIGURE 3.22 Targeted microring lens profile (line curve) and measured 1D ring lens profile data (dotted curve). (a) Ring lens fabricated from intensity-based phase-grating mask. (b) Ring lens fabricated from resist profile-based phase-grating mask. (From Sung, J., et al., *J. Microlith. Microfab. Microsyst.*, 4, 041603-1, 2005. With permission.)

profile-based phase-grating design is in good agreement with the target spherical surface, while the microlens fabricated with the intensity profile-based phase-grating design shows a significant deviation from the targeted spherical surface. The reason for this mismatch between the designed and fabricated lens surfaces is that the transmittance-based phase-grating design does not account for the nonlinear response of the resist to the exposure light. But, the excellent agreement between the fabricated and designed lens profile in case of the resist profile-based phase-grating mask demonstrates that our numerical convolution concept for predicting the analog resist profile with a phase-grating mask works accurately. We also measured the ring lens surface profiles made from two different designs of phase-grating masks. Since the ring lens surface does not fit to the aspheric sag function of radius, only the 1D height profile data was acquired from the Zygo surface scanning and compared with the target design profile as shown in Figure 3.22. Again, it shows clearly that the resist profile-based phase-grating mask better matches the targeted ring lens profile. The scanning electron microscope (SEM) picture of fabricated ring lens is shown in Figure 3.23.

3.5 CONCLUSION

Photolithographic steppers are a proven toolset for the fabrication of micro-optical elements at the wafer level. Although numerous methods exist for analog intensity profiles, the phase mask approach has some advantages over other methods in the

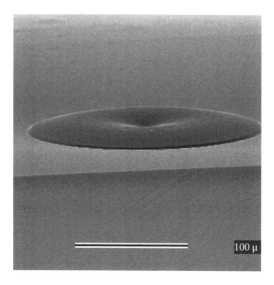

FIGURE 3.23 SEM picture of the 200 μm ring lens made from resist-based phase-grating mask. (From Sung, J., et al., *J. Microlith. Microfab. Microsyst.*, 4, 041603-1, 2005. With permission.)

ease of fabrication and integration in standard processes. The ability to fabricate phase masks is quite common at most mask houses and the features are not beyond what is possible on both laser writers and e-beam lithography tools.

In the fabrication of the phase-grating mask for analog elements, the control of line-width and pixel size was very critical for making the exact analog surface as designed. There are a few parameters affecting the profile such as the e-beam dose/current, developing time for PMMA e-beam resist, and the initial thickness of PMMA resist. We went through experimental tuning of those parameters to make a phase mask with exact duty cycle distribution as designed. If there is significant error in the duty cycle of the actual phase-grating mask, it will result in an incorrect resist height profile since the transmittance of the phase-grating mask will deviate from the desired pattern.

In order to maintain a good repeatability of analog resist profiles from wafer to wafer, the process should be tightly controlled. The resist and developer should always be in good condition and should be used well before its expiration date. Appropriate resist processing should be followed, such as the baking condition and the sitting time prior to each process step. Also the stepper system should be maintained in good condition to ensure consistent exposure results. This requirement on process control is quite standard for lithographic processing making it simple to maintain. Thus, this analog resist profile fabrication process using the phase-grating mask is easily transferable to product manufacturing environment.

REFERENCES

1. W. Henke, W. Hoppe, H.J. Quenzer, P. Staudt-Fischbach, and B. Wagner, Simulation and experimental study of gray-tone lithography for the fabrication of arbitrary shaped surfaces, In *Micro Electro Mechanical Systems 1994, MEMS '94, Proceedings, IEEE Workshop*, January 25–28, 1994, pp. 205–210.
2. W. Daschner, R. Stein, P. Long, C. Wu, and S.H. Lee, One-step lithography for mass production of multilevel diffractive optical elements using high energy beam sensitive (HEBS) gray-level mask, *Proc. SPIE: Diffr Hologr Opt Technol III*, vol. 2689, 1996, pp. 153–155.
3. B. Morgan and C.M. Waits, Development of deep silicon phase Fresnel lens using gray-scale lithography and deep reactive ion etching, *IEEE J. Microelectromech. Syst.*, 13 (1), February 2004, 113–120.
4. C.M. Waits, R. Ghodssi, M.H. Ervin, and M. Dubey, MEMS based gray-scale lithography, In *Semiconductor Device Research Symposium, 2001 International*, December 5–7, 2001, pp. 182–185; J. Shimada, O. Ohguchi, and R. Sawada, Microlens fabricated by the planar process, *J. Lightwave Technol.*, 9, May 1991, 571–576.
5. E.-H. Park, M.-J. Kim, and Y.-S. Kwon, Microlens for efficient coupling between LED and optical fiber, *IEEE Photon. Technol. Lett.*, 11, April 1999, 439–441.
6. C. David, B. Nohammer, and E. Ziegler, Wavelength tunable diffractive transmission lenses for hard X-rays, *Appl. Phys. Lett.*, 79 (8), August 2001, 1088–1090.
7. M. Pitchumani, H. Hockel, W. Mohammed, and E.G. Johnson, Additive lithography for fabrication of diffractive optics, *Appl. Opt.*, 41 (29), 10 October 2002.
8. M. Levinson, N. Viswanathan, and R. Simpson, Improved resolution in photolithography with a phase-shifting mask, *IEEE Trans. Electron. Dev.*, 29 (12), December 1982, 1812–1846.
9. A. Kwok-Kit Wong, *Resolution Enhancement Techniques in Optical Lithography*, vol. TT47, SPIE Press, Bellingham, WA, 2001, Chapters 2, 3, and 5.
10. J. Sung, H. Hockel, and E.G. Johnson, Analog micro-optics fabrication using two dimensional binary phase-grating mask, *Opt. Lett.*, 30 (2), January 2005, 150–152.
11. J. Sung, H. Hockel, J. Brown, and E.G. Johnson, Refractive micro-optics fabrication with a 1-D binary phase-grating mask applicable to MOEMS processing, *J. Microlith., Microfab., Microsyst.*, 4 (4), October–December 2005, 041603-1.
12. J. Sung, H. Hockel, J.D. Brown, and E.G. Johnson, Development of a two-dimensional phase-grating mask for fabrication of an analog-resist profile, *Appl. Opt.*, 45 (1), January 2006, 33–43.

4 Electron Beam Lithography for the Nanofabrication of Optical Devices

Aaron Gin and Joel R. Wendt

CONTENTS

The ability to fabricate optical elements with submicron critical features has enabled the creation of true integrated optical microsystems. Refractive and diffractive optics, micromirrors, etc., have been created in a variety of materials, such as fused silica, III–V materials and silicon. As a result, the dreams of "lab-on-a-chip," snapshot polarimetry, and other compact optical systems have been realized.[1,2] The most critical fabrication steps in microoptical systems include electron beam (e-beam) lithography direct writing and pattern transfer using subtractive or additive processes. This chapter describes the practical fabrication challenges involved in creating optical devices with critical dimensions of less than a micrometer. Material-specific e-beam lithography and processing techniques and several device case studies are also presented.

4.1 INTRODUCTION

Using state-of-the-art fabrication technologies, optical engineers have been able to realize microoptical devices in a large variety of materials (semiconductors, glass, plastics, etc.). These micro- or nanoscale elements offer optical flexibility not possible with traditional, bulk optical components. Furthermore, the size of these devices allows multiple functionalities on the same substrate as well as easier integration with conventional electronics and control systems. A wide assortment of optical devices has been used for various applications such as diffractive or refractive microoptics, micromirrors, integrated waveguides, near-field optics, and microprisms.[3]

E-beam lithography is an enabling research tool for nanoscience and engineering due to its powerful ability to write nanometer-scale features on a variety of substrate materials. The capability to pattern fine structures in a deterministic, repeatable fashion is desirable in virtually every field of science. When coupled with state-of-the-art clean room and semiconductor device fabrication facilities, the power of the e-beam lithography tool is multiplied. Thus, the tool operator must have a strong understanding of the starting materials and the entire fabrication process in order to choose the appropriate e-beam lithography parameters that will be compatible with the overall process and result in sufficiently high yield. Important decisions such as resist tone, type, thickness as well as pattern "bias" or offset and other adjustments must be made in order to correctly write the pattern in the e-beam resist. Finally, proper pattern transfer with metal deposition and liftoff, etching, additive techniques, etc., is critical to realizing the full potential of the e-beam system.

When applied to optical devices, e-beam lithography and nanofabrication techniques can be used to produce novel optical components (such as micropolarizers and waveplates) as well as compact optical systems (micro total analytical systems,[4] lab-on-a-chip and polarimeters). It is not sufficient to simply transfer known good processes (from silicon, for instance) to optical devices and materials. In this chapter, we explore the material and device-specific fabrication challenges encountered during the processing of modern microoptical devices.

4.2 ELECTRON BEAM LITHOGRAPHY

4.2.1 Electron Beam Lithography Overview

The advantages of e-beam lithography include the ability to form features much smaller than those possible using optical techniques, which are fundamentally limited by the wavelength of light used. Depending on the application, e-beam spot sizes used in lithography range from a few nanometers up to hundreds of nanometers. Note that the deBroglie wavelength[5] of an electron is given by

$$L = \frac{1.2}{\sqrt{V_b}} \text{ nm}$$

where V_b is the acceleration voltage. Given that most e-beam systems operate at 1 kV or more, the electron wavelength has no impact on limiting the resolution achievable with e-beam lithography.

E-beam lithography systems are variously used for photolithography mask creation, advanced prototyping and nanoscale scientific research and development. At ultimate resolution, state-of-the-art e-beam lithography systems can enable pattern transfer (with metal liftoff, etch, or additive techniques) of critical dimensions below 10 nm. Furthermore, these sorts of patterns can be realized in a variety of different materials, such as semiconductors like silicon and GaAs, insulators such as fused silica, amorphous diamond, SiO_2 and SiN as well as various metals.

Conventional e-beam lithography involves the deterministic scanning of a highly focused beam of electrons across a surface with electron-sensitive resist. Various positive- and negative-tone resists are used in e-beam lithography and are discussed below. In high-resolution lithography systems, a thermal field emission source such as a ZrO/W emitter is often used to create the stream of electrons. Several stages of electrostatic and/or magnetic lenses help focus and shape the beam within the electron column and electromagnetic deflection coils are used to scan the beam across the working field (typically 100 μm–1 mm on a side depending on the system and beam parameters). Dedicated e-beam lithography systems typically rely on a laser-interferometric stage to move the sample between the exposures of different working fields. By counting interference fringes as the stage moves, stitching errors or the offset errors between adjacent write fields, can be kept to a minimum (less than 20 nm). A set of electrostatic beam blanking plates is used to deflect the e-beam out of the electron optical path when needed.

A computer system controls the deflector coils to write the pattern and the blanker to regulate the amount of electron dose incident on a particular pixel. The simple area dose formula can be written as

Dose × area = beam current × dwell time = total number of incident electrons

One of the biggest disadvantages of e-beam lithography, relatively long write times from the serial nature of the exposure process, can be observed by plugging in some numbers for dose, exposed area, and beam current. A high-resolution beam current might be around 500 pA, and assuming 300 nm thick 495 K poly(methyl methacrylate) (PMMA) (~1000 μC/cm² for an isolated line at 50 kV acceleration voltage), 25 MHz pattern generator, and 10 nm pixel size, the time to write a 1 cm² area (50% coverage) would be about 5.5 h. Once you add in stage movement time and beam "settle" times, the wait for a beam to be stable between deflections, writing large areas can obviously be extremely time-consuming.

E-beam lithography patterns can be aligned to existing features on the sample by using registration or alignment marks. These can be deposited or etched features generally simple geometric shapes or cross-shaped marks. The features must allow high-contrast backscatter electron imaging, and therefore are usually formed from metal (Au for instance) or etched features. The operator then prepares the job file to find and recognize the existing marks on the substrate surface to determine and compensate for rotation, gain, and position errors. Typical registration errors on state-of-the-art e-beam lithography systems can be less than 20 nm.

4.2.2 Electron Beam Lithography Systems

There are several major manufacturers of dedicated e-beam lithography systems including JEOL, Vistec, and Elionix. These vendors offer systems with a range of capabilities depending on the need, such as CD/DVD mastering, high throughput, and research and development tools. The state-of-the-art research-grade Gaussian-spot, vector-scan tool today features 50 or 100 kV acceleration voltage with minimum beam spot sizes below 5 nm. Beam position with these systems can now be controlled below 2 Å according to specifications from the newest 100 kV tools.

There are also other e-beam lithography tools with aftermarket pattern generators (such as the Nabity NPGS) on commercial scanning electron microscope (SEM) systems. These systems rely on a combination of add-on hardware and software to control the e-beam and stage.

Today, e-beam lithography systems have beam spot sizes on the order of single nanometers in diameter, but resolution in developed e-beam resist is limited by additional factors. Forward and backward electron scattering effects can significantly broaden linewidths written in resist. And any given resist has an ultimate resolution limit set by the molecule size of its constituents.

The work reported in this chapter was performed on two dedicated e-beam lithography systems, a JEOL JBX-9300FS and a JEOL JBX-5FE. These tools have 100 and 50 kV maximum acceleration voltage, respectively, and both utilize a thermal field emission source.

4.2.3 Electron Beam Lithography Techniques

4.2.3.1 Electron Beam Resists

E-beam resist selection varies depending on application and processing restrictions. First, an e-beam operator must choose the resist tone that will be used in a particular exposure. A positive-tone resist may be employed if the pattern to be transferred is fairly sparse, for instance the thin lines of an optical waveplate device. Negative-tone resist might be used if a relatively large area of the resist needs to be developed away. This could occur for instance if the resist was being used as an etch mask in a process. A few positive- and negative-tone electron-sensitive resists and their attributes are described here.

PMMA, which is also known as Plexiglas or acrylic, was one of the first materials to be used as a positive-tone e-beam resist and it remains the standard to which all other e-beam resists are measured. It exhibits very high resolution with long shelf life. Its drawbacks include the need for relatively high electron doses and little or no resistance to most dry etch chemistries.

PMMA is usually found in two molecular weights for e-beam lithography including 495 K and 950 K and is solvent-diluted in either anisole or chlorobenzene for spin casting. After resist spinning, PMMA is typically hot-plate baked from about 15 min up to several hours at ~170°C. When exposed with high energy electrons, PMMA undergoes chain scission, which allows chemical development. A mixture of 1:3 methylisobutylketone (MIBK):isopropanol (IPA) is typically used to achieve high-contrast, low sensitivity development. Under practical conditions, PMMA can achieve developed resist features on the order of 15 nm in resist thickness less than 100 nm. The ultimate resolution of PMMA has been shown to be about 10 nm for thin resist exposed on a membrane to reduce backscatter.

ZEP-520A is a high-resolution positive copolymer resist and is several times more sensitive than PMMA. ZEP forms a film ~3000 Å thick when spun at approximately 6000 RPM. The resist can then be baked at ~170°C for typically 3 min. After exposure, development is performed using xylene or *n*-amyl acetate (for high resolution).[6]

NEB-31A3 is a negative-tone resist that can achieve relatively high resolutions (sub-50 nm isolated linewidths) for a negative resist. Coating at 4000 RPM results in a 300 nm film thickness. The sample is softbaked at 100°C for 2 min. The typical electron dose is also about an order of magnitude less than PMMA. After exposure, a hot-plate bake at 90°C is applied to the sample for 1 min. Development is performed using MF-321, which is 21% tetramethyl ammonium hydroxide (TMAH) with some proprietary surfactants.

4.2.3.2 Charge Dissipation Layers

Insulating substrate materials have traditionally made e-beam lithography challenging due to electron charging issues.[7] This occurs due to a build up of negative charges in the localized area near the e-beam spot. In order to reduce these charging effects, researchers have introduced a conductive charge dissipation or "flash" layer to conduct electrons away from the write area.[8,9] Highly accelerated electrons can travel through the thin metal layer to expose the resist and localized charge is quickly dissipated away through the conductive layer.

At Sandia National Laboratories, the traditional charge dissipation layer is a thin (5–10 nm) Au film that is thermally deposited after the e-beam resist (typically PMMA) is spun and baked. Following a conventional e-beam lithography exposure, the metal layer is etched away with a potassium iodide/iodine solution.[10] The sample can then be developed in a normal manner. This method allows significant flexibility over other charge dissipation techniques, such as those involving a metal layer underneath the e-beam resist. Some charge dissipation techniques include the deposition of a metal layer before coating the sample with resist. Although this method works, it comes with the added inconvenience of transferring the developed resist pattern into the metal layer with an anisotropic metal etch. Further issues include the fact that the metal may not be compatible with the component application, for instance, if the device is to be used in optical transmission.

Relatively new ways to create charge dissipation layers include the application of a water-soluble conducting polymer solution. Products like ESpacer from Showa Denko present e-beam lithography operators with another way to deal with substrate charging due to insulating materials. Issues here include high cost, limited availability, and short shelf life.

4.2.3.3 Proximity Effect

Proximity effect arises when large features are adjacent to small areas such as nano-scale lines or dots, or when patterns are densely packed.[11] In these situations, the electron dose is not confined solely to the intended pattern geometry primarily due to electron scattering effects. This nonuniformity can lead to round corners meant to be right angles as well as increase linewidth and diameter for nanoscale lines and dots, respectively. There are many ways to compensate for these proximity effects,[12,13] but the most practical method is often to assign different doses to critical features.

By using care in the pattern design, dose compensation can be achieved by proper CAD file design (taking advantage of graphic data system (GDS) layers and data-types, for instance) which allows users to define multiple shot ranks (different doses for different datatypes). For instance, to properly expose a small 50 nm line that expands in width to a bond pad (~100 μm on a side), it is often useful to identify the thin line, taper and bond pads with different datatypes and assign larger electron doses to smaller features. This is typically termed "dose modulation."

Furthermore, it is common practice to introduce "bias" to a pattern, generally narrowing the CAD geometry to compensate for proximity effect. In this case, the operator can bias the pattern file by several tens of nanometers to more readily real-ize the intended pattern shape and size in developed resist.

4.2.3.4 Other Electron Beam Techniques

By spinning on a high molecular weight PMMA (or other less-sensitive resist) on top of a more sensitive lower layer of PMMA, undercut profiles can be achieved after exposure and development. This sort of resist profile can be useful to achieve "shad-owed" metal evaporations as well as improved metal liftoff processes.

Another bilayer technique involves applying the inverse technique—applying the higher sensitivity resist on top of the lower sensitivity resist. Metallization with this technique can form a "T-gate" structure. Researchers have recently formed these features using a ZEP520A and UVIII bilayer.[14]

Analog lithography involves the use of graduated electron doses to partially develop away various depths of electron resist. As shown in Figure 4.1, three-dimensional structures can be fabricated in resist using this technique. Analog methods require careful calibration of electron dose versus developed resist depth to achieve repeatable

Vacc.: 100 kV
Beam current: 400 pA
Resist: ZEP520 1 μm
Shot ranks: 250
(8 nm step)

JEOL

FIGURE 4.1   Three-dimensional resist structures formed by graduated electron dose con-trol. (Image courtesy of JEOL.)

results. Furthermore, low-contrast e-beam resists and/or low-contrast developers are required to achieve reasonable smoothness in the developed patterns.

4.3 MATERIAL-SPECIFIC OPTICAL DEVICE NANOFABRICATION

4.3.1 Overview

A variety of materials are useful for optical devices due to their respective properties that may include absorption, reflection, electro-optic coefficient, and emission characteristics. The device material system is generally chosen with applications in mind. Processing of these devices can become very complex, because optical materials can range from metals to semiconductors to insulators depending on the application need. This difficulty is compounded by the fact that many state-of-the-art optical devices require subwavelength pattern transfer. This section will discuss a few of the most common optical materials in use today as well as selected material-dependent fabrication techniques.

4.3.2 Silicon

Silicon is one of the most studied materials in the world because of its prominence as a standard material within integrated circuits. It offers high temperature semiconductor operation, and has an easily grown oxide insulator material. These features, as well as all the existing industrial infrastructure and tooling, make it attractive to develop integrated silicon optical devices. Silicon is used in optoelectronic devices such as light-emitting diodes (LEDs),[15,16] waveguides and solar cells,[17] as well as in micromachined electromechanical systems.[18]

Micro- and nanofabrication methods in silicon are fairly well established to enable the realization of these optical devices. The classic anisotropic wet etch chemistry for silicon is $KOH:H_2O$ (1:1 by weight) at 80°C.[19] Etch rates of approximately 1.4 µm/min have been observed. This chemistry is often used in silicon micromachining to achieve via holes or to release membranes. Anisotropic dry etching of silicon has been achieved with SF_6 chemistries.[20] Subsequent research into very high aspect ratio dry etching resulted in deep reactive ion etch (DRIE) techniques with alternating SF_6/Ar and CHF_3/Ar chemistries.[21] The latter gas mixture promotes the deposition of a Teflon-like polymer on all exposed surfaces. The SF_6/Ar step then quickly removes the polymer from the horizontal surfaces due to ion bombardment and etches the exposed silicon surface. This polymer step prevents sidewall etching and results in a highly anisotropic profile.[22]

E-beam lithography has been demonstrated on silicon substrates with PMMA linewidths below 10 nm.[23] Although not discussed in the case studies below, Sandia National Laboratories has fabricated many optical devices in silicon including surface-micromachined mirror devices[24] and concentrator solar cells.[25]

4.3.3 Gallium Arsenide

Gallium arsenide (GaAs) is a standard semiconductor material used in various electronic and optoelectronic devices. Due to the intrinsic properties of GaAs, high frequency operation, electron mobility, and light emission characteristics are generally superior to similar measures in silicon. Engineers have taken advantage of these characteristics to make a wide variety of optoelectronic devices such as solar cells,[26] detectors,[27] and quantum well lasers.[28,29] GaAs is transparent over the entire 2–5 μm band and has a high refractive index. A drawback to gallium arsenide with respect to silicon is a lack of a robust native oxide. Due to advances in GaAs device fabrication for high-speed and high-power applications, material processing details are relatively well known.

GaAs works well with e-beam lithography as charge transport in the material is generally adequate, obviating the need for a charge dissipation layer. Isolated resist trench linewidth in PMMA can readily be demonstrated below 20 nm.[30]

Wet etching of GaAs is achieved by several oxidizers including peroxides and halogens such as bromine. Anisotropic etching along the (111) planes can be achieved with a $H_2PO_4:H_2O_2:H_2O$ (3:1:50).[31] One suggested selective etchant system is citric acid:$H_2O_2:H_2O$ (5 g:2 mL:5 mL) that offers a 10–100:1 selectivity between AlGaAs and GaAs and can offer sacrificial etches of epitaxial layers, for example.

Other researchers have optimized a selective, anisotropic etch for GaAs with an AlGaAs etch stop.[32] The inductively coupled plasma (ICP) system used a $BCl_3:SF_6:N_2:He$ chemistry and was more than 200 times more selective for GaAs than AlGaAs. This process also produced excellent sidewall passivation on GaAs with a very high degree of anisotropy.

At Sandia National Laboratories, a chemically assisted ion beam dry etch (CAIBE) process has been developed to achieve highly anisotropic features in GaAs for a waveplate device described in the GaAs waveplate work below. The CAIBE system uses an argon ion beam to control the physical energy and flow rate of the ambient reactive gas at the sample surface. In this case, the ambient reactive gases were a combination of Cl_2 and BCl_3. Etch conditions were optimized to maximize sidewall anisotropy and trench depth.

4.3.4 Fused Silica

Fused silica is a type of glass containing silicon dioxide (SiO_2) in an amorphous form. Due to a high-purity synthesis process, optical and thermal properties of fused silica are generally superior to other types of glass. Fused silica has a very low coefficient of thermal expansion (~5.5 × 10⁻⁷ cm/(cm·K) from 20°C to 320°C), making it useful for environments with extreme and/or varying temperatures. Because of its high transparency from ultraviolet to mid-infrared and its low refractive index (and associated low Fresnel reflection losses), fused silica is often viewed as an ideal optical material. Applications of this material consist of various microfluidics devices[33] as well as optical elements for a wide range of wavelengths.[34]

In order to process this material using modern clean room tools, various techniques specific to the properties of fused silica have been developed. Previous work has centered on wet etching using hydrofluoric acid and wafer bonding for microfluidic devices.[35] Due to the fact that fused silica is an amorphous material, isotropic profiles are observed during wet chemical etching. Using photolithography and buffered hydrofluoric acid, nearly perfect half-cylinders can be formed in fused silica. Subsequent wafer bonding techniques (at ~1000°C) can form cylindrical microfluidics channels.

Modern nanofabricated optical devices can require step or angled features to realize Fresnel lenses or other diffractive elements, such as binary, gradient index,

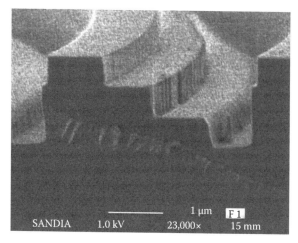

FIGURE 4.2   SEM showing the step profiles generated in fused silica for diffractive optical devices. (From Kemme, S.A. et al., *Proc. SPIE.*, 5347, 247, 2004.)

or subwavelength artificial gradient index lenses. At Sandia National Laboratories, researchers use several rounds of e-beam lithography and reactive ion etching (RIE) to achieve binary lenses with multilevel step profiles in fused silica. Combining a Ni hard mask (with Cr or Ti adhesion layer) with an optimized etch recipe of 40 sccm trifluoromethane (CHF_3), 3 sccm O_2 with 200 W RF power at 40 mTorr and 396 V susceptor bias, researchers realized very anisotropic features in this material as shown in Figure 4.2.[36]

The starting material quality of fused silica can play an important role in the final condition of processed devices. Low-quality fused silica can exhibit voiding and surface roughness, or grass, after RIE steps. For this reason, it is usually important to select high-quality substrates in order to achieve the best possible anisotropy and surface smoothness—traits which are almost universally desirable in optical elements.

4.4   OPTICAL DEVICE FABRICATION CASE STUDIES

4.4.1   Fused Silica SEED Device

Recent work on high-speed optical computing devices at Sandia National Laboratories has focused, in part, on a fused silica diffractive optical element (DOE) set that performs the optical interconnect function among symmetric self-electro-optic effect devices (S-SEED).[36] Two variations of DOE devices were designed for use with incident wavelengths of either 860 or 1550 nm. Both DOEs involved two e-beam lithography mask layers, which enable four levels in the fused silica optical material. Blazed lens arrays were designed and translated into an optical/e-beam lithography mask set. An algorithm was written that mapped each aspherical lens phase profile into this four-level, stepped phase profile, wrapping every two *pi* radians. A second algorithm was developed to assign the appropriate phase steps to a corresponding mask level for the fabrication process. Critical variables in this design process were the feature size, etch depth, and duty cycle. A plan view of the interface between two of the 100% fill-factor DOEs is shown in Figure 4.3.

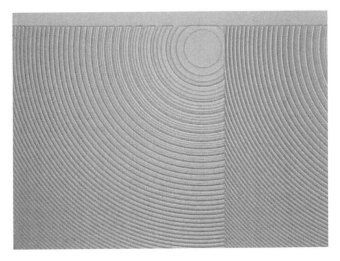

FIGURE 4.3   Optical microscope image of a portion of a DOE for optical interconnects.

The fabrication steps required to realize this device are similar to those published elsewhere.[37] The major steps include

1. Solvent clean 0.9 mm thick fused silica wafer
2. Metal alignment marks
    a. After Hexamethyldisilazane (HMDS), spin on 6% 495 K PMMA to ~5000 Å
    b. Flash 75 Å Au for JEOL e-beam write and develop
    c. Evaporate 100 Å of Cr then 2000 Å of Au
    d. Liftoff
3. First etch metal mask
    a. Repeat step 2 with 4% 495 K PMMA to ~3300 Å. Metal mask is 100 Å of Cr and 500 Å of Ni
4. First etch photoresist
    a. After HMDS, spin on AZ4330 to ~1.3 μm
    b. Bake at 90°C for 2 min
    c. Expose 8 s
    d. Develop in AZ400 K 1:4
5. First etch in RIE Plasmatherm 790
    a. 40 sccm CHF_3, 3 sccm O_2, at 200 W RF for ~4800 Å etch depth
    b. Strip metal mask in HCl for 3 min and solvent clean
6. Second etch metal mask
    a. Repeat step 2 with 9% 495 K PMMA to ~20,000 Å. Metal mask is 100 Å Cr and 850 Å Ni
7. Second etch photoresist
    a. Repeat step 4
8. Second etch in RIE Plasmatherm 790
    a. Repeat step 5 for etch depth of ~9600 Å

After a solvent clean, the fused silica wafer is e-beam written to put down thick metal alignment marks in Cr/Au. Another e-beam write is used to pattern the first etch mask in Cr/Ni and photoresist is patterned to cover the large regions away from the DOEs. The sample is etched with a RIE system and $CHF_3:O_2$ chemistry to a depth of about 480 nm. The metal mask is then stripped with hydrochloric acid. Thicker PMMA resist is spun to cover the etched topography and an e-beam write is performed to pattern another Cr/Ni etch mask. After another patterned photoresist protective layer, the sample is etched to a depth of 960 nm.

The e-beam lithography steps are complicated by the fact that fused silica is an insulating material. As mentioned before, a charge dissipation layer is applied above the e-beam resist. The concentrated scanning of alignment marks can be problematic due to resist swelling and outgassing. This can cause the Au charge control layer to crack, degrading the charge dissipation properties and introducing spurious scan signals. In general, the best way to avoid these problems is to image and scan registration marks as quickly as possible, using the minimum number of machine scans to achieve proper registration.

Additionally, successive e-beam lithography steps require careful dose calibrations that depend upon the resist topography. For instance, e-beam operators must take into consideration the resist thickness to cover relatively large steps in material and properly expose varying thicknesses of resist, depending on the spatial position of the dose. An example of alignment error is shown in Figure 4.4. This misalignment is most likely due to a bad or degraded mark that the e-beam machine references.

Finally, the etch transfer of the pattern may be complicated by aspect-ratio dependent etching (ARDE) effects,[38] which represent a difference in etch rate that depends on the amount of exposed material and local topography. ARDE is dependent upon the transport of ions and neutrals in the microstructure. The transport is affected by the angular distribution for ions as well as their incident energy. With many microoptical devices having subwavelength dimensions and varying spatial frequency features these aspect-ratio dependent effects are frequently observed. Thus, the dry

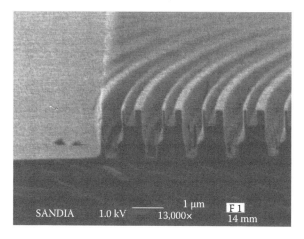

FIGURE 4.4   SEM image of fused silica diffractive element with ~50 nm lithography registration error.

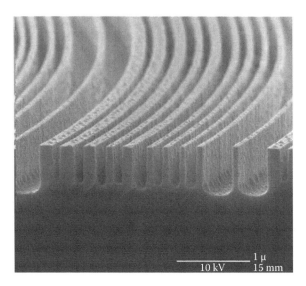

FIGURE 4.5   SEM image of GaAs ARDE.

etch transfer involved in this project needed several rounds of process iteration to optimize the recipe. ARDE effects are particularly prominent in various blazed sub-wavelength gratings due to extreme differences in the aspect ratios of various structures. Figure 4.5 shows an example of ARDE effects on a GaAs grating with varying pitch and width.[39]

4.4.2   FUSED SILICA MICROPOLARIZER

Imaging polarimeters collect the polarization dependence of light from a given scene and are useful in applications such as remote sensing. In order to process transient polarimetric information, it is necessary to simultaneously collect data sufficient to calculate the four Stokes parameters. These parameters form the complete Stokes vector and can be used to describe the polarization state of electromagnetic radiation. This simultaneous data collection method is called snapshot polarimetry and requires all polarization orientations (and the waveplate described in the next section) to be present in an array of superpixels.

Linear wire-grid polarizers are perhaps the most basic subwavelength diffractive optic component. Incident light polarized perpendicularly to the grid (transverse magnetic, TM) is transmitted efficiently, while transverse electric (TE) light is primarily absorbed and reflected. For devices operating from the visible to mid-wave infrared (MIR) wavelength regime (3–5 μm), e-beam lithography, and pattern transfer techniques become important tools for the fabrication of these devices. Using these methods, we are able to control the pitch, depth, and duty cycle down to the tens of nanometer range. In the Sandia National Laboratories design, a focal plane array super pixel, which consists of four pixels with different polarization angles, was fabricated as shown in Figure 4.6.[40]

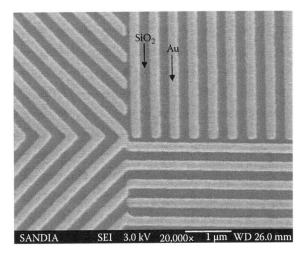

FIGURE 4.6   SEM image of center of micropolarizer pattern illustrating the four metal grating angles.

We utilize subwavelength metal gratings that result in an effective index that is dependent upon the incident polarization. The micropolarizer devices were fabricated on fused silica wafers with Ti/Au metal. Major fabrication steps included

1. Solvent clean ~0.5 mm thick fused silicon wafer
2. After HMDS, spin on 495 K C4 PMMA to ~3000 Å
3. Flash thermal Au (75 Å)
4. JEOL e-beam exposure
5. Etch charge dissipation layer with KI:I solution
6. Pattern development in 1:3 MIBK:IPA
7. Evaporate 100 Å of Ti, then 1500 Å of Au
8. Metal liftoff in acetone
9. Solvent clean

The wire-grid polarizer devices were fabricated with a single e-beam lithography write and Ti/Au metal liftoff. Due to the insulating nature of fused silica, a thin gold flash layer is deposited before exposure to dissipate charge. This metal layer is removed after e-beam exposure with a potassium iodide:iodine wet etch.

The e-beam written width was 0.15 μm with a bias reduction of 25% to produce the desired duty cycle of 50% or a linewidth of 0.2 μm and period of 0.4 μm. To allow a range of duty cycles, several patterns were exposed on a single substrate with varying electron dose. This approach resulted in devices with duty cycles that varied from 44% to 55%. The 48% duty cycle devices were ultimately used in the experimental device verification. A SEM image of the optimized polarizer device is shown in Figure 4.7.

Metal grating devices with varying aperture sizes were used to investigate the extinction ratio of transmitted light. A polarizer with 18 × 18 μm aperture is shown

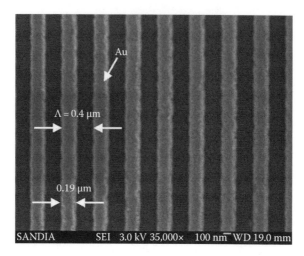

FIGURE 4.7   SEM of polarizer section with gold lines that have a 48% duty cycle.

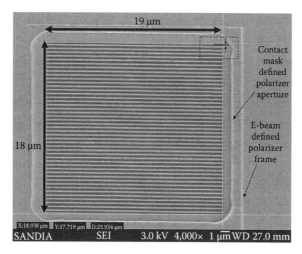

FIGURE 4.8   SEM image of a test micropolarizer device. Device has 45 gold wires with 400 nm period, 50% duty cycle, and 150 nm thickness. This micropolarizer has a termination edge space of about 1/4 wavelength.

in Figure 4.8. Extinction ratio quantifies the polarizer efficiency for TM wave transmission versus TE wave transmission. These microdevices averaged extinction ratios around 100:1, which exceeded the previous best wire-grid polarizer performance by a factor of 7.[41]

4.4.3   GaAs Birefringent Waveplate Device

A critical component for collecting the complete Stokes vector in polarimetry is the waveplate or phase retarder that alters the incident light polarization as it is transmitted through the waveplate.

Researchers at Sandia National Laboratories have fabricated a birefringent diffractive waveplate array in gallium arsenide[42] in order to retard the incident radiation by a quarter wave or 90° over the 2–5 μm waveband. GaAs was chosen for its large form birefringence (different effective index of refraction depending upon the polarization of incident light), transparency in this spectral window, and ease of processing. The fabrication steps proceeded as follows:

1. Solvent clean ~0.5 mm thick GaAs
2. Deposit 30 nm thick SiO_2
3. After HMDS, spin on 4% 495 K PMMA to ~2500 Å
4. E-beam exposure and development
5. Exposed SiO_2 is etched with RIE
6. Etch GaAs trenches with CAIBE
7. Wet etch with buffered HF to remove SiO_2

A thin layer of SiO_2 was deposited on the GaAs surface to act as an etch mask for the GaAs material. Afterwards, e-beam resist was applied and the e-beam lithography exposure of PMMA and development were conducted. A fluorine-based RIE was utilized to remove the exposed SiO_2 features. The aforementioned CAIBE process was used to anisotropically deep etch the exposed GaAs and the SiO_2 etch mask was removed with a buffered HF dip.

Initial processing runs resulted in duty cycles approximately 10% smaller than desired. Subsequent lithography attempts approximated the target duty cycle almost exactly. The rectangular groove shape obtained was nearly identical to the as-designed structure. The optimized etched ribs had a period of 650 nm, duty cycle of about 80% and a trench depth of 1.23 μm as shown in Figure 4.9.

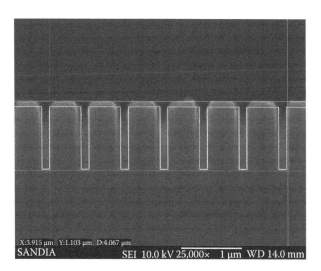

FIGURE 4.9 Fabricated waveplate device before removal of oxide etch mask. The white line shows the design target.

4.5 CONCLUSION

The realization of optical devices using e-beam lithography and other nanofabrication techniques has been discussed in this chapter. In particular, we have described the processing of an optical interconnect for all-optical computing and two elements (a micropolarizer and waveplate) for an imaging polarimeter system at Sandia National Laboratories. The most critical fabrication steps of these devices have been detailed including e-beam lithography and pattern transfer. The demonstrated adaptation of well-known nanolithography and fabrication techniques to nontraditional materials like fused silica and GaAs has led to greater flexibility in optical device design and application. Further advances in materials and device fabrication will undoubtedly drive the state of the art in optical systems with nanometer-scale minimum feature size.

ACKNOWLEDGMENTS

Sandia is a multiprogram laboratory operated by Sandia Corporation, a Lockheed Martin Company, for the United States Department of Energy's National Nuclear Security Administration under contract DE-AC04-94AL85000.

REFERENCES

1. Herzig, H.P., *Micro-Optics*. Taylor & Francis, London, U.K., 1997.
2. Sweatt, W.C., Warren, M.E., Kemme, S.A., Christenson, T.R., Blum, O., Sinclair, M.B., Butler, M.A. and Gentry, S.M., A plethora of micro-optical systems. *Proc. SPIE*. 4437, 125–133, 2001.
3. Sinzinger, S. and Jahns, J., *Microoptics*. Wiley-VCH Verlag, Weinheim, 1999.
4. Vilkner, T., Janasek, D., and Manz, A., Micro total analysis systems. Recent developments. *Anal. Chem.* 76, 3373–3386, 2004.
5. McCord, M.A. and Rooks, M.J., Electron beam lithography. *SPIE Handbook of Microlithography, Micromachining and Microfabrication*, Volume 1 (Microlithography), Rai-Choudhury, P. (ed.). SPIE Publications, London, U.K., 1997.
6. Matsui, S., Nanostructure fabrication using electron beam and its application to nanometer devices. *Proc. IEEE* 85(4), 629–643, 1997.
7. Liu, W., Ingino, J., and Pease, R.F., Resist charging in electron beam lithography. *J. Vac. Sci. Technol. B*. 13(5), 1979–1983, 1995.
8. Mancini, D.P., Gehoski, K.A., Ainley, E., Nordquist, K.J., Resnick, D.J., Bailey, T.C., Sreenivasan, S.V., Ekerdt, J.G., and Willson, C.G., Hydrogen silsesquioxane for direct electron-beam patterning of step and flash imprint lithography templates. *J. Vac. Sci. Technol. B*. 20(6), 2896–2901, 2002.
9. Aktary, M., Jensen, M.O., Westra, K.L., Brett, M.J., and Freeman, M.R., High-resolution pattern generation using the epoxy novolak SU-8 2000 resist by electron beam lithography. *J. Vac. Sci. Technol. B*. 21(4), L5–L7, 2003.
10. Wendt, J.R., Warren, M.E., Sweatt, W.C., Bailey, C.G., Matzke, C.M., Arnold, D.W., Allerman, A.A., Carter, T.R., Asbill, R.E., and Samora, S., Fabrication of high performance microlenses for an integrated capillary channel electrochromatograph with fluorescence detection. *J. Vac. Sci. Technol. B*. 17(6), 3252–3255, 1999.
11. Chang, T.H.P., Proximity effect in electron-beam lithography. *J. Vac. Sci. Technol.* 12(6), 1271–1275, 1975.

12. Owen, G. and Rissman, P., Proximity effect correction for electron beam lithography by equalization of background dose. *J. Appl. Phys.* 54(6), 3573–3581, 1983.

13. Parikh, M., Corrections to proximity effects in electron beam lithography. II. Implementation. *J. Appl. Phys.* 50(6), 4378–4382, 1979.

14. Chen, Y., MacIntyre, D., and Thoms, S., T-gate fabrication using a ZEP520A/UVIII bilayer. *Microelec. Eng.* 57–58, 939–943, 2001.

15. Zimmermann, H., *Integrated Silicon Opto-Electronics*. Springer, Berlin, 2000.

16. Ng, W.L., Lourenco, M.A., Gwilliam, R.M., Ledain, S., Shao, G., and Homewood, K.P., An efficient room-temperature silicon-based light-emitting diode. *Nature* 410, 192–194, 2001.

17. Wong, H., Recent developments in silicon optoelectronic devices. *Microelec. Reliab.* 42, 317–326, 2002.

18. Elwenspoek, M. and Jansen, H.V., *Silicon Micromachining* (Cambridge Studies in Semiconductor Physics and Microelectronic Engineering). Cambridge University Press, Cambridge, U.K., 1999.

19. Madou, M.J., *Fundamentals of Microfabrication: the Science of Miniaturization*, 2nd edn. CRC Press, Boca Raton, FL, 2002.

20. Kovacs, G.T.A., Maluf, N.I., and Peterson, K.E., Bulk micromachining of silicon. *Proc. IEEE* 86(8), 1536–1551, 1998.

21. Yeh, J.L.A., Jiang, H., and Tien, N.C., Integrated polysilicon and DRIE bulk silicon micromachining for an electrostatic torsional actuator. *JMEMS* 8(4), 456–465, 1999.

22. Laermer, F. and Schilp, A., Method of anisotropically etching silicon, US Patent Number 5501893.

23. Vieu, C., Carcenac, F., Pepin, A., Chen, Y., Mejias, M., Lebib, A., Manin-Ferlazzo, L., Couraud, L., and Launois, H. Electron beam lithography: Resolution limits and applications. *Appl. Surf. Sci.* 164(1–4), 111–117, 2000.

24. Hetherington, D.L. and Sniegowski, J.J., Improved polysilicon surface-micromachined micromirror devices using chemical mechanical polishing. *Proc. SPIE.* 3440, 148–153, 1997.

25. Sinton, R.A., Kwark, Y., Gan, J.Y., and Swanson, R.M., 27.5-Percent silicon concentrator solar cells. *IEEE Elect. Dev. Lett.* EDL-7(10), 567–569, 1996.

26. Bertness, K.A., Kurtz, S.R., Friedman, D.J., Kibber, A.E., Kramer, C., and Olson, J.M., 29.5%-efficient GaInP/GaAs tandem solar cells. *Appl. Phys. Lett.* 65(8), 989–991, 1994.

27. Hasnain, G., Levine, B.F., Bethea, C.G., Logan, R.A., Walker, J., and Malik, R.J., GaAs/AlGaAs multiquantum well infrared detector arrays using etched gratings. *Appl. Phys. Lett.* 54(25), 2515–2517, 1989.

28. Kumar, S., Williams, B.S., Hu, Q., and Reno, J.L., 1.9 THz quantum-cascade laser with one-well injector. *Appl. Phys. Lett.* 88(12), 121–123, 2006.

29. Bank, S.R., Wistey, M.A., Yuen, H.B., Goddard, L.L., Ha, W., and Harris J.S. Jr., Low-threshold CW GaInNAsSb/GaAs laser at 1.49 µm. *Elec. Lett.* 39(20), 1445–1446, 2003.

30. Dial, O., Cheng, C.C., and Scherer, A., Fabrication of high-density nanostructures by electron beam lithography. *J. Vac. Sci. Tech. B.* 16(6), 3887–3890, 1998.

31. Tong, M., Ballegeer, D.G., Ketterson, A., Roan, E.J., Cheng, K.Y., and Adesida, I., A comparative study of wet and dry selective etching processes for GaAs/AlGaAs/InGaAs pseudomorphic MODFETs. *J. Electr. Mater.* 21(1), 9–15, 1992.

32. Lee, J.W., Devre, M.W., Reelfs, B.H., Johnson, D., Sasserath, J.N., Clayton, F., Hays, D., and Pearton, S.J., Advanced selective dry etching of GaAs/AlGaAs in high density inductively coupled plasmas. *J. Vac. Sci. Technol. A* 18(4), 1220–1224, 2000.

33. Kovacs, G.T.A., *Micromachined Transducers Sourcebook*. WCB McGraw-Hill, Boston, MA, 1998.

34. Malitson, I.H., Interspecimen comparison of the refractive index of fused silica. *J. Opt. Soc. Am.* 55(10), 1205–1209, 1965.

35. Grosse, A., Grewe, M., and Fouckhardt, H., Deep wet etching of fused silica glass for hollow capillary optical leaky waveguides in microfluidics devices. *J. Micromech. Microeng.* 11, 257–262, 2001.

36. Kemme, S.A., Peters, D.W., Carter, T.R., Samora, S., Ruby, D.S., and Zaidi, S.H., Inadvertent and intentional subwavelength surface texture on microoptical components. *Proc. SPIE.* 5347, 247–254, 2004.

37. Wendt, J.R., Krygowski, T.W., Vawter, G.A., Blum, O., Sweatt, W.C., Warren, M.E., and Reyes, D., Fabrication of diffractive optical elements for an integrated compact optical microelectromechanical system laser scanner. *J. Vac. Sci. Technol. B* 18(6), 3608–3611, 2000.

38. Gottscho, R.A., Jurgensen, C.W., and Vitkavage, D.J., Microscopic uniformity in plasma etching. *J. Vac. Sci. Technol. B.* 10(5), 2133–2147, 1992.

39. Wendt, J.R., Vawter, G.A., Smith, R.E., and Warren, M.E., Subwavelength, binary lenses at infrared wavelengths. *J. Vac. Sci. Technol. B.* 15(6), 2946–2949, 1997.

40. Kemme, S.A., Cruz-Cabrera, A.A., Nandy, P., Boye, R.R., Wendt, J.R., Carter, T.R., and Samora, S., Micropolarizer arrays in the MWIR for snapshot polarimetric imaging. *Proc. SPIE.* 6556, 655–604, 2007.

41. Kemme, S.A., Snapshot polarimetry enables new signature opportunities for remote sensing. *SPIE Newsroom*, DOI: 10.1117/2.1200709.0861, 2007.

42. Boye, R.R., Kemme, S.A., Cruz-Cabrera, A.A., Vawter, G.A., Alford, C.R., Carter, T.R., and Samora, S., Fabrication and measurement of wideband achromatic waveplates for the mid-infrared region using subwavelength features. *J. Microlith. Microfab. Microsyst.* 5(4), 043007, 2006.

5 Nanoimprint Lithography and Device Applications

Jian (Jim) Wang

CONTENTS

Based on mechanical replication, nanoimprint lithography (NIL) is an emerging technology that can achieve lithographic resolutions beyond the limitations set by light diffractions or beam scatterings in conventional lithographic techniques, while promising high-throughput patterning. This chapter reviews the status and some of the recent progress in the commercial applications of this technology.

5.1   INTRODUCTION

Lithography is the enabling technology for patterning various types of micro- and nanostructures in applications that include integrated circuits (ICs), microelectromechanical systems (MEMS), patterned media, biochips, and photonic and optoelectronic devices. The ability to make patterns at micro- to nanoscale is crucially important to the advance of modern science and technology. Until now the patterning process has largely taken place through optical lithography, which involves illuminating a photomask and imaging the resultant pattern to expose a photoresist-coated wafer.

As the demand for device performance, density, and speed increases, critical dimensions (CD) of the patterned features drop. According to the International Technology Roadmap for Semiconductors (ITRS), the semiconductor industry is pushing to reduce the transistor gate length down to 22 nm by 2010. The challenges of following such a roadmap are multifold, the biggest of which is believed to be lithography. The cost of a single next generation lithography tool will exceed $50 millions in the next few years—a formidable price tag for most potential users.

In recent years, researchers have investigated a number of alternative and potentially low-cost lithography techniques including soft lithography (e.g., microcontact printing), AFM/dip-pen lithography, and NIL [1–6]. Of these three techniques, the most notable and successful one so far is NIL. In NIL and its variants, including thermal nanoimprint (thermal-NIL) [1,2], UV-nanoimprint (UV-NIL) [3–6], step-and-repeat nanoimprint (e.g., step-and-flash imprint lithography (S-FIL) [4]), resist patterning is done nontraditionally by deforming mechanically the resist materials. The mechanical replication nature of NIL removes the resolution-limit factors (such as light diffraction and e-beam scattering) that are often inherent with the traditional approaches.

The purpose of this chapter is to introduce and review the most recent progress of NIL and its applications. We focus primarily on commenting on and reviewing the process, technology, and applications, all from the commercial applications' point of view. (The reader is referred to two other recent reviews [7,8] on NIL for more research-oriented discussions.)

5.2   HISTORY OF IMPRINT PATTERNING AND IMPRINT LITHOGRAPHY

The principle of NIL is quite simple. NIL uses a mold to form a material mechanically (i.e., resist), at its soft and/or liquid state, into a solid and negative replicate of the mold surface's patterns. In order to do a follow-on pattern transfer, the replicated resist layer on top of the wafer needs to be thin, in a dimension similar to the size of the pattern features. Depending on the applied resist materials, NIL can be categorized into thermal-NIL or UV-NIL. For the thermal-NIL, the resist is a spin-on thermoplastics thin film such as poly(methyl methacrylate) (PMMA) [1,2]. During nanoimprinting, the resist layer is first heated above its glass transition temperature, T_g, to lower its viscosity. Next, the mold prints into the resist layer to replicate the patterns, and then the resist layer is cooled below its T_g before the mold is removed. In contrast, the UV-NIL [3–6] starts with a photopolymerizable resin, which is typically a low-viscosity liquid. The mold replicates the patterns using the combined effect of mechanical embossing and liquid capillary action followed by a UV polymerization process that is required to solidify the resist patterns before the mold is removed. The division between the thermal-NIL and the UV-NIL can be blurred if a mixed approach is used, while a polymerizable resist can be solidified through either a thermal polymerization process or a combined thermal/UV polymerization process.

The earliest idea of using the mechanical imprint to do patterning (but not necessarily lithography) can probably trace back hundreds if not thousands of years. Perhaps, the earliest proposal to use the mechanical imprint for microlithography was published in a US Patent (US 4,035,226) [9] by A. S. Farber and J. Hilibrand of RCA Corporation on July 12, 1977 (the patent was initially filed on April 14, 1975). What the patent discussed was actually a thermal-NIL process based on a moldable, thermoplastic resist layer. In another US patent (US 4,731,155) [10] by L. S. Napoli and J. P. Russell of General Electric Company, published on March 15, 1988, a similar thermal-NIL was discussed, and an example of patterning 0.7 micron features was demonstrated. In the US patent 4,512,848 [11] published on April 23, 1985, a lithography process based on generating an intermediate transfer mask (by casting a master), a process which we call today "reverse imprint lithography," was proposed. In [11], a patterning resolution of 100–200 nm was demonstrated by using a 3600 lines/mm holographic grating as the master. To the author's best knowledge, the earliest UV-NIL was published, in the US patent 5,259,926 [12] by K. Kazuhiro et al. of Hitachi Ltd., on November 9, 1993.

The imprint lithography work mentioned above somehow did not receive much attention until 1995, when Chou et al. demonstrated a 25 nm resolution in their milestone work on NIL [1]. Follow-up work [13–17] by Chou's group on NIL eventually turned NIL into a red-hot field. The S-FIL work in 1999 [4,18] further accelerated the acceptance of NIL because it provided a feasible solution for fine alignment between two patterned layers. This alignment was crucial for making integrated microelectronic devices.

5.3   VARIOUS ASPECTS OF NIL

5.3.1   NANOIMPRINT COMPONENTS AND PROCESSES

5.3.1.1   Patterning Resolution

Patterning resolution is definitely the number one advantage of NIL. In the original work by Stephen Chou et al. [1], 25 nm diameter vias were demonstrated with a thermal nanoimprinting into PMMA. Shortly after that, sub-10 nm vias [2] were successfully demonstrated. The molds used in the above work were made by e-beam lithography.

Recently, using a photocurable NIL (UV-NIL) process [19], Chou's Princeton group successfully produced lines of the polymer resist just 6 nm wide with a pitch of only 12 nm. To produce the nanoimprint mold with 6 nm features [19], authors created a superlattice of alternating layers of GaAs and AlGaAs by using molecular-beam epitaxy (MBE), and then selectively etched away the AlGaAs layers with hydrofluoric acid. Thus far, the patterning resolution study for nanoimprinting has been limited by technology for the fabrication of molds with even smaller features as also for accurate measurement of such small features. Today's e-beam lithography struggles to create patterning with a pitch smaller than 30 nm.

5.3.1.2   Resist

The imprint resist is the core part of the NIL process. Typically, the resist formula is a trade secret for all the commercial nanoimprint players. Each nanoimprint tool manufacturing has its own resist formula, which is developed either by the tool manufacturing itself (e.g., Obducat (www.obducat.com/) and Nanonex (www.Nanonex.com)) or through its partners (e.g., Molecular Imprint (www.molecularimprints.com/), Suss Microtec (www.suss.com), and EV group (www.evgroup.com/en)). Molecular Imprint Inc. (MII), for example, partners with the University of Texas at Austin and Brewer Science for resist development and production. EV group has partnered with the Micro Resist Technology of Germany on the resist.

For the thermal-NIL, it is desirable to do patterning with a low viscosity state during the imprinting process, as well as to do the imprint at a lower temperature. This desire naturally leads to using low T_g thermoplastics as resists. However, another important criteria for a good NIL resist is that the final resist pattern should maintain its mechanical integrity during the mold–wafer separation as well as during subsequent pattern transfer steps (at room temperature or even at elevated process temperature). Although a number of very low T_g materials can be found, most of them may not meet the mechanical integrity requirement. In addition, other requirements, such as spin-coating and reactive ion-etching performance, also need to be considered. So far, the most successful thermal-NIL resists have typical process temperatures around 80°C–100°C. Higher than 100°C is not preferred or necessary, but lower than 80°C typically encounters a resist structure stability issue.

A further desire on the resist for lower imprint temperatures naturally leads to the consideration of UV-curable and/or thermal curable/thermosetting chemicals. For the UV-NIL resist, the requirements are: (1) low viscosity for easy flow and pattern forming; (2) fast UV curing speed; (3) easy mold separation and no sticking; (4) good thermal, mechanical, and plasma etching (e.g., RIE) properties after curing; (5) easy cleaning from mold if any sticking occurs due to particles or defects; and (6) spin-coating capability with good adhesion and uniformity.

Recently, NanoOpto Corporation reported a UV-curable and single-layer spin-coating-capable resist [6], which has shown excellent performance, while meeting all of the above requirements. The spin-coat-capable, UV-curable resist consists of the following compositions: (a) monomers (diluents) and oligomer resins (including (methyl)acrylates and epoxies, polyether, polyester, polysiloxane, and polyurethane); (b) photoinitiators and photosensors (free-radical or cationic species, such as darocure 1173 and irgacure 369); (c) plasticizers, such as butyl octyl phthalate; and (d) other compositions (e.g., internal release agents, compatibilizer, lubricants and other stabilizers, such as fluorinated- or siloxane-based structures). The UV-curable resist used in this work consists of (methyl)acrylates, polysiloxane, irgacure 369, butyl octyl phthalate, and siloxane. In addition to the single-layer spin-coating performance, one big advantage of the NanoOpto's UV resist is that it is O_2 etchable, the same as most conventional photoresists.

The resist of the S-FIL [18] process has a very low viscosity (~4 cps) but a high organosilicon content. It cannot be etched by oxygen plasma but requires fluorine etching chemistry. This complicates the final etching mask-removal process after postlithography pattern transfer processes. A bilayer resist structure is required to resolve this issue.

5.3.1.3　Resist Residual Layer

One of the unique characteristics of NIL is the existence of a thin so-called "residual layer" of the resist in the compressed area of patterns. Due to the viscous nature of the resist at the imprint, this residual layer is inevitable. Because the residual layer needs to be etched off in order to expose the underneath layer for further process, the existence of this residual layer has a significant implication. The thicker the residual layer, the more the etching effort required for removing the residual layer; this could potentially lead to a CD change and variation due to the lateral attack to resist lines during the etching. Therefore, for applications (such as in semiconductors) which require a very tight control on CD, a thinner residual layer is always preferred. Various simulation efforts were conducted to assist in the understanding of the resist flow dynamics at imprinting. In general, a low-resist viscosity at the imprinting stage is required in order to achieve a thin residual layer. Additionally, the amount of the resist applied or the spin-coat thickness and the imprint pressure also affect the thickness of the residual layer.

A thickness of 30–50 nm (or more) was typically reported. In Figure 5.1, Obducat achieved a 10 nm uniform residual resist layer with its nanoimprint resist and process. Very recently, Hitachi also reported [20] a 10 nm residual resist layer across a 65 mm diameter wafer with a thermal-NIL approach.

5.3.1.4　Nanoimprint Molds/Templates and Mold Treatment

The fabrication of the mold, or the so-called template, is the key technology to the success of NIL. Because of the typical 4× projection imaging reduction factor, the state-of-the-art DUV lithography needs only ~250 nm resolution to make its photomasks to print 65 nm features. For NIL, a template technology with 1× resolution is required. The existing industrial infrastructure for supporting DUV photomasks by

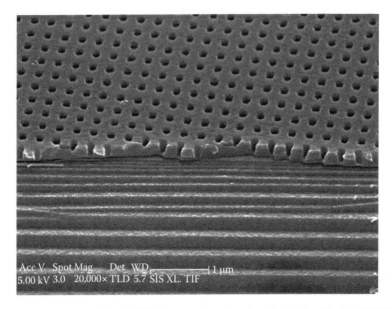

FIGURE 5.1 Cross-sectional SEM image of an imprint obtained with the 100/300 nm stamp. The residual layer thickness is homogeneously 10 nm. (Courtesy of Mr. Ken Mason of Obducat AB.)

an e-beam and/or a laser beam scanning writer does not offer such high resolution. In order to achieve a broad industrial acceptance of NIL, a new infrastructure for the 1× template fabrication, inspection, [21] and repair [22] needs to be established. This is quite challenging. Obducat uses its own e-beam lithography technology to support and develop its mold technology [23]. Molecular Imprint works with many industrial partners, including Photronics Inc., Toppan Photomasks (template fabrication), Motorola, KLA-Tencor (template inspection), and Carl Zeiss (template repair), toward establishing the template infrastructure. Recently, BenchMark Technologies, Inc. started offering e-beam-written standard nanoimprint test molds.

In order to avoid the resist sticking to the mold during separation, the mold surface needs to be treated with a mold-release coating. The most widely used technique is based on surface modification with self-assembled monolayers (SAM). SAM is typically applied by a solvent immersion or vapor treatment technique. Vapor treatment, a technique which has been widely used in semiconductor fabrication, is more suitable for industrial manufacturing. One mold release agent published is 1H, 1H, 2H, 2H-perfluorodecyl-trichlorosilane (FDTS) [6]. PVD and/or CVD deposition of teflon-like thin coating has also been used.

5.3.1.5 Different Forms of NIL: Pros and Cons

5.3.1.5.1 Thermal-NIL vs. UV-NIL

The advantage of using the UV-curable resist vs. thermoplastics is that it avoids the thermal cycling process, a process which should not be used if alignment/overlay is required. Thermal distortion, along with the mismatch of the thermal expansion coefficients between the mold and the wafer, prevents thermal-NIL from achieving alignment/overlay accuracy well below the submicron level. In addition, the UV curing process is typically much faster than the thermal process due to slow heating and cooling cycles in the thermal process. The UV curing process normally takes only a few seconds, as opposed to a few minutes or longer as in the thermal process.

The major issues with the UV-NIL are volume shrinkage at UV curing due to the phase transition from liquid to solid (this problem was reported in [3]) and the difficulty in forming a uniform thin film from the UV-curable resist by spin coating. Due to the second issue, most UV-NILs, including S-FIL, utilize either a spread-droplet approach [4] or a bilayer resist approach to form a thin film [24]. The spread-droplet approach [4,18] applies one or multiple droplets of the UV-curable resist onto the wafer, and then spreads the droplet over a thin film while engaging a mold toward the wafer. This approach could potentially cause cross-wafer variation of the resist residual layer due to the existence of any mechanical nonuniformity. The spread-droplet approach could also potentially generate unwanted bead edges around the mold edge because of the technical challenge of applying the exact amount of resist. To get the same residual layer thickness, the amount of resist applied needs to be pattern-density-dependent. The bilayer resist approach requires coating of one additional buffer polymer layer between the UV-curable patterning layer and the substrate. This buffer polymer layer works as an adhesion/wetting layer for the UV-resist layer. As a result, an extra etching process is required during the pattern transfer.

5.3.1.5.2 Step and Repeat vs. Full-Wafer-Imprinting

For the semiconductor microelectronics (i.e., CMOS) application, layer to layer overlay/alignment accuracy of sub-20 nm is required for patterning 50 nm features. A full-wafer-imprint approach is unlikely to achieve this requirement, particularly for large wafers. Therefore, a step-and-repeat process, such as the S-FIL, has to be adopted to meet the ultimate overlay accuracy.

For most optical- and bioapplications (as opposed to semiconductor applications), highly precise alignment and registration with accuracy down to the submicron or nanometer scale may not be required [6]. Rather, large area high-throughput and low-cost patterning are the keys. Therefore, for most optical applications, full-wafer patterning may be preferred more than the step-and-repeat approach. For applications that require a large chip size ($\gg$30 mm) with unstitched patterns, such as large-area optical nanogratings, full-wafer patterning is a must.

Hitachi recently reported [20] full-wafer patterning of 300 mm diameter wafers with thermal-NIL, where good uniformity was achieved.

5.3.1.6 Defect

The defect issue, due to the mask and wafer contact, can be traced back to the 1960s and 1970s from contact photolithography. The defect for imprint lithography [6,25] is even more of an issue than the contact photolithography, due to its imprint nature. To meet commercial quality manufacturing requirements, defect was identified as the top issue for commercial applications of NIL. As NIL moves into the product manufacturing stage, the impact of defect on the product quality and yield receives more and more attention [25].

For NIL, defects can roughly be divided into two groups: random distributed defects and repeated defects. Random distributed defects include particles, incomplete contact during the imprint, and the residues after the imprint, which are not repeatable in terms of location and amount. Repeated defects include the existing defects on the mold and the substrate which are repeated in the process. We can further divide the random defects into particle-associated defects (PADs), gap (or void)-associated defects (GADs), and separation-related defects.

Clearly, PADs are not like the particles in DUV photolithography which only generates defects close to the particle size: PADs are particle-amplified defects. The NIL is extremely sensitive to any particle existing between the wafer and the mold during the imprinting process. Because a thinner resist layer is required to pattern smaller features, the smaller the pattern feature size, the more sensitive the NIL process is to the particles. The particles could also damage the mold surface, which can be very costly. The size of the impact area is related to the particle size, hardness, substrate and mold stiffness, applied pressure, and polymer surface properties. A detailed analysis on the relation of the particle size to the final defect size can be found in reference [25].

Besides PAD, NIL presents another unique defect which is associated with the gap between the mold and the resist during the imprinting process. GAD is generated by wafer/mold irregularities or a bad process control, such as vacuum and pressure. Unlike PAD, which has a clear physical image, GAD has no clear boundary and is not obvious to visual inspection. GADs normally are incomplete pattern structures with variable sizes. The incomplete pattern structures are due to the polymer shortage when filling the gap between the mold and the substrate. According to the previous study, the surface charge on the mold tends to pull the polymer up, if not contacted, to form unique polymer patterns. GADs are process-related defects which initially do not exist on the substrate or the mold. The origination of the gap could be the wafer/mold bending, surface waving, or vacuum and pressure control. A quantitative analysis showed that both wafer bending and surface waving or roughness can lead to formation of the gap during contact printing.

It is believed that cleaning technologies for the mold and the substrate are crucial for the commercial success of NIL [25]. Some work has been reported along this line [6,25]. Those technologies fall into some trade secrets of companies.

5.3.1.7 Alignment and Overlay

Although alignment/overlay was once considered to be one of few top challenges for NIL, significant progress has been made in this area over the past few years. Compared to a ~1 micron alignment accuracy demonstrated by Zhang and Chou [26] in 2001, a sub-500 nm overlay accuracy was demonstrated by Molecular Imprint two years ago with the IMPRIO 100 machine, and most recently, a sub-50 nm overlay accuracy has been achieved by Molecular Imprint (machine IMPRIO 250). The alignment result from MII is based on a 26 × 33 mm field size for the step-and-repeat operation. With a recently announced Suss MicroTec's NPS 300 nanopatterning stepper, alignment with a 250 nm accuracy was specified with an imprint field size of up to 100 mm.

5.3.1.8 Throughput

Wafer patterning throughput is an important factor for modern semiconductor manufacturing if market requirements are to be met. A state-of-the-art DUV stepper typically is capable of processing 100 wafers (8"–12") per hour.

For MIIs IMPRIO machines, the current throughput is about 5 wafers (8"–12")/h. In each S-FILs field imprint, there are several subprocesses, which include resist applying, mold engagement, alignment, UV shining for curing, and demolding. In comparison, in each DUV field print in a DUV stepper, only alignment and UV shining steps are needed. Due to these inherent differences, it is expected that S-FIL will have a lower throughput than the modern DUV stepper. It is unclear at this moment how far the throughput can be pushed for the S-FIL nanoimprint process. 40–50 wafers/h could be the entry criteria for using nanoimprint for CMOS production.

In the full-wafer NIL, a throughput of more than 10 wafers/h has been achieved without considering a separate demolding process [6]. For full-wafer nanoimprinting, depending on the wafer size, the demolding process could be a time-consuming process as well. No published data exists on this parameter.

5.3.1.9 Separation (Step and Repeat vs. Whole Wafer)

Mold-wafer separation, or so-called demolding after imprinting, is another unique nature of the nanoimprint process that differs from conventional DUV photolithography. The separation process is crucial for any commercial high-volume manufacturing using nanoimprint. To ease separation and avoid resist peeling, the mold has to be coated with a mold release layer. In addition, a certain mold release material is typically incorporated into the resist material to further ease the demolding process. The demolding force can be highly dependent on the density, depth, aspect ratio (i.e., surface area), and geometry of the patterns on the mold. A mold with densely packed and deep/high-aspect-ratio grating lines is most difficult to demold.

For a full-wafer NIL, most demolding processes were done manually and separately from the nanoimprint process. How to integrate the demolding process into the full-wafer thermal-NIL process is still a technical challenge. This is because the demolding force could be significantly larger than the vacuuming force between the mold (or substrate) and the vacuum chuck holding them. The bigger the wafer size, the more challenging the demolding process will be. With a large wafer, the wafer and/or mold could be bent or deformed during the demolding process. This creates an additional problem in the mass production of precise nanostructure patterns.

In comparison, because of a large area ratio between the wafer and the mold, the step-and-repeat NIL process (such as S-FIL) has a big advantage from the demolding point of view. With a 300 mm diameter wafer and a 25 × 25 mm mold, the area ratio is 113. Assuming that a mold can be fixed onto a top fixture with a special process, the vacuum chuck will be able to hold the large wafer during the demolding process with the "amplification" factor of 113.

5.3.1.10 Three-Dimensional Imprint

Another unique feature of NIL is its capability of three-dimensional (3D) patterning, and this could be the second most important advantage of nanoimprint following the resolution. Although photolithography can also do some gray-scale patterning, NIL

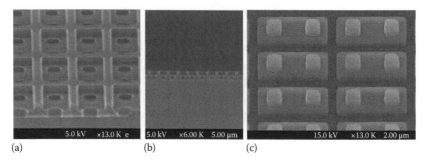

(a)   (b)   (c)

FIGURE 5.2 Top (a) and side (b) views of imprinted 3D nanostructures with a multitier mold (c). (Courtesy of Molecular Imprint.)

is much more flexible and naturally suitable for doing various 3D nanopatternings with high resolution.

Figure 5.2 shows patterned via chain structures by S-FIL with a multitiered mold (template) designed for metal interconnects in CMOS fabrication. In the current CMOS fabrication by DUV lithography, the metal and via layers require separate lithographic steps. Using nanoimprint lithography, this can be done with one single lithography step. Halving the number of lithography steps for the interconnect layers can be significant as the number of interconnect layers increases.

5.3.2 Nanoimprint Machines

5.3.2.1 Obducat

Obducat AB has sold more nanoimprint machines than anyone else. More than 50 Obducat NIL machines (two models) have been deployed to date. The laboratory version (NIL2.5) accepts a substrate size up to 65 mm in diameter, while the industrial version (NIL6) accepts a substrate size up to 100 mm in diameter. Both models are thermal nanoimprint machines with the UV-module as an option. Low-end Obducat machines are priced in the ~$200 K range. Figure 5.3a (left) shows the NIL6 nanoimprint machine made by Obducat.

5.3.2.2 Molecular Imprint

Based on the S-FIL™ developed at the University of Texas at Austin by Prof. G. Willson's group [4], MII offers several different models of nanoimprint machines: Imprio 55 for entry-level systems; Imprio 100 for pilot-line systems; and Imprio 250 for advanced litho systems. All three systems offer a sub-50 nm resolution. Imprio 55 can do overlay down to 1 micron, while Imprio 100 and 250 can do overlay down to 500 and 50 nm. All three systems can handle up to 200 mm wafers, while Imprio 250 (Figure 5.3b) can also handle up to 300 mm. Imprio 55 is priced at ~$600 K, while Imprio 250 is priced at ~$5–6 M, with a fully automated template and wafer loading (i.e., cassette to cassette operation) and automated alignment capabilities. So far, the Imprio 250 is the only imprint machine which includes cassette to cassette operation as its standard feature, possibly because this is the first machine ever truly designed for meeting automated production needs.

FIGURE 5.3 (a) An Obducat nanoimprint machine NIL6 (upper left), (b) an IMPRIO 250 machine made by Molecular Imprint (upper right), (c) a Nanonex NX-2500 full-wafer imprinter (lower left), (d) and a EV group EVG-620 NIL system (lower right).

The core technologies for MII include a precise resist applying method through a MEMS-controlled nozzle, automated alignment/overlay control, and template leveling mechanics. In addition, MII is based on a low-viscosity resist, which allows the achievement of a thin residual resist layer.

5.3.2.3 Suss MicroTec

In cooperation with the VTT Microelectronics Center, Suss MicroTec announced its new nanopatterning stepper, NPS 300, in the spring of 2005. Optimized for cost-effective production of nanostructures, the NPS 300 is able to combine, on the same platform, aligned hot (thermal) and cold (UV) nanoimprinting. The NPS 300 is able to print sub-50 nm geometries with an overlay accuracy of 250 nm. The NPS 300 is a flexible tool, and is available either as a manually loaded machine or a fully automated system for hands-off operation. The latter configuration includes fully automated wafer handling for sizes up to 300 mm and an automated template pickup capability that allows different templates to be printed on the same wafer. It can accept templates sized up to 100 mm and thickness up to 6.5 mm, and substrates with a size up to 300 mm and thickness up to 8 mm. The NPS 300 tool is priced at ~$700 K–800 K.

5.3.2.4   EV Group

EVG (EV Group) offers nanoimprint tools based on their previously existing tools and platforms. Since 1995 or even earlier, EVG offered wafer bonding tools for the semiconductor and photonics industry. Its wafer bonder, which is capable of nanoimprinting, can bond a silicon wafer with a glass wafer under well-controlled conditions of temperature and pressure. EVG also provides an upgrade service with a specially designed fixture for UV nanoimprinting to its long existing mask aligners, allowing customers to do UV nanoimprinting with previously purchased mask aligners. The add-on feature costs much less than any of the nanoimprint machines.

5.3.2.5   Nanonex

Chou's group at the University of Minnesota, and later at Princeton University, was the first to develop nanoimprint machines since 1995. Nanonex has the longest machine development cycle because of its tie with Chou's group. The uniqueness of Nanonex's commercialized machines is that they use gas rather than plates to apply pressure for nanoimprinting [27]. The advantage of gas pressure is that it is isostatic: the resulting force uniformly pushes the mold and the resist-coated substrate together, and shear and rotational components are minimized. Moreover, since the mold and/or the wafer are flexible rather than rigid, conformation between the mold and the film is achieved regardless of unavoidable deviations from planarity. All the other nanoimprint machines use a plate to engage the mold and the substrate. To improve pressure uniformity, either a soft "cushion" layer is inserted between the plate and the mold/wafer, or special attention is paid to the flatness of the plate, the mold, and the wafer. The major disadvantage of the gas imprint is that sealing the mold/wafer edge is required: this adds complexity to the system design and also complicates and even jeopardizes the alignment process. The gas imprint also allows using lamps to heat the resist layer, which leads to fast thermal-NIL cycles.

5.3.2.6   Other Homemade Systems

Few other application-oriented companies develop nanoimprinting machines internally, primarily because the early-stage nanoimprint machines did not offer many technical advantages over what the application-oriented companies were able to do internally.

Wang et al. [6] reported a UV nanoimprint machine developed by NanoOpto Corporation for the production of optical devices. The machine was designed by using a commercial, high-pressure piston press system. The bottom plate of the piston press is a quartz window which can transmit UV light from a high-power UV lamp for curing purposes. The center part of the press system is enclosed in a small vacuum chamber. The mold and the resist-coated wafer are loaded into the press machine vacuum chamber with spacers between them to prevent them from contacting each other. Since the coated and baked resist layers are in a liquid state, any contact between the mold and the coated wafer could lead to air enclosure, which would form trapped air bubbles and defects after imprinting and curing. The spacers have a separation distance of about 25 microns. After completely vacuuming the air between the mold and the wafer, the spacers are removed and the mold and the wafer are pressed together with a uniformly distributed pressure from the press system. To improve the pressure uniformity, cushion layers such as rubber films are used between the press plates and the mold/substrate. The cushion layers help to distribute pressure evenly across the whole imprinted area. While holding the pressure, the resist is cured by a shining UV light through the bottom window plate and the transparent substrate. The system is capable of delivering a pressure up to 400 PSI (for a 4" wafer), depending on the process requirement. The curing process only takes about 5 s. After the imprint, the mold is separated from the wafer. Good imprint uniformity is obtained across a 4" wafer.

Hitachi also developed its own thermal nanoimprint machine [20]. It includes alignment, transfer, and press subprocess portions. The machine is capable of imprinting up to 300 mm diameter wafers. Recently, they reported [20] results of a full 300 mm diameter wafer thermal nanoimprint.

5.4 COMMERCIAL DEVICE APPLICATIONS

In this section, we discuss some of the most important progress made in the last two years in applying NIL to commercial device fabrication and its applications. These represent the first round of commercially oriented applications, including applications in optics, photonics, display, molecular electronics, and data storage areas.

5.4.1 NEAR-IR POLARIZERS FOR TELECOMMUNICATION

NanoOpto has been using the full-wafer UV-nanoimprint, based on a single-layer spin-coat resist, in the pilot production of high-performance, near-IR (telecom) nanowire-grid polarizers and polarizing beam splitters since 2002 [6,28–34]. Figure 5.4a is a SEM photograph of a cross-sectional view of a part of the high-aspect-ratio (aspect ratio: the height divided by the width) nanogratings from a 4" wafer made by the UV-NIL. Figure 5.4b shows a finished near-IR polarizer after metal shadowing and nanotrench filling. The metal nanogratings are buried into dielectric materials to ensure the best environmental stability and reliability. Broadband (e.g., from 1260 to 1610 nm) performance with a high efficiency (>97.5% for transmission of the polarized light and >97.5% for reflection of the polarized light) and a high extinction ratio (>40 dB for transmission and >20 dB for reflection) has been achieved in volume production [6,34]. In the past three years, hundreds of 4" in-diameter wafers for the nano-optic polarizers have been produced by UV-NIL. In Figure 5.5, performance data of the probability distribution function (PDF) based on 20 nano-optic polarizer wafers of 4" in-diameter are shown for the transmission extinction ratio. This indicates a good fabrication consistency of performance from wafer to wafer.

5.4.2 VISIBLE POLARIZERS FOR PROJECTION DISPLAY

NIL has also been used to fabricate high-performance visible nanowire-grid polarizers for display applications (i.e., projection display). LG electronics reported [35] 50 nm

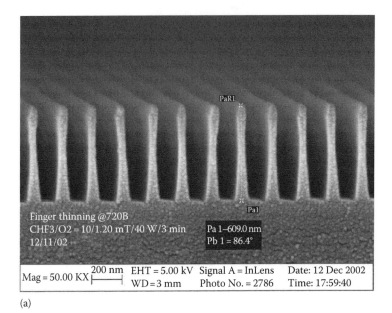

Finger thinning @720B
CHF3/O2 = 10/1.20 mT/40 W/3 min
12/11/02

PaR1

Pa1

Pa 1–609.0 nm
Pb 1 = 86.4°

Mag = 50.00 KX   200 nm   EHT = 5.00 kV   Signal A = InLens   Date: 12 Dec 2002
                            WD = 3 mm   Photo No. = 2786   Time: 17:59:40

(a)

Cursor height = 594.6 nm

Cursor width = 37.82 nm

P05BS 6412
after trenchtill

200 nm   EHT = 7.00 kV   Signal A = InLens   Date: 13 Feb 2005   LEO
          WD = 7 mm   Photo No. = 1444   Time: 17:39:25

(b)

FIGURE 5.4 SEM of a high-aspect-ratio SiO_2 grating and a near-IR telecom polarizer fabricated by the UV-NIL.

half-pitch aluminum nanowire-grid polarizers with a 2000:1 extinction ratio and an 83% transmittance at a wavelength of 470 nm. Figure 5.6 shows a 2″ × 2″ square of a visible polarizer made by LG electronics with the thermal NIL.

NanoOpto Corporation reported [36,37] both the high-extinction ratio and high-transmission versions of visible polarizers made by UV-NIL. Figure 5.6 shows a 60 nm wide and 130 nm tall aluminum nanowire polarizer. Broadband operation

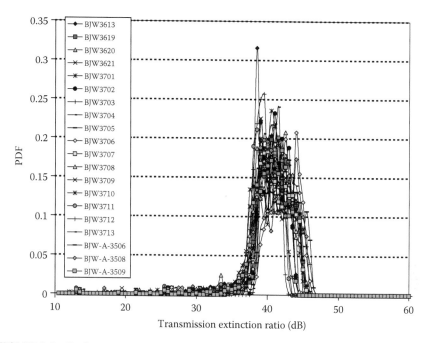

FIGURE 5.5 Performance distribution for the transmission extinction based on results from 20 processed 4" in-diameter nano-optic polarizer wafers. PDF is the probability distribution function based on on-wafer optical performance mapping data.

from 400 nm to >1.55 microns, along with a 4" diameter, was reported. Polarizers with a transmission of >85% (with >23 dB extinction ratio) and an extinction ratio of >50 dB (with >60% transmission) were achieved.

5.4.3 Optical Waveplates for Optical Pickup Units (CD/DVD)

Figure 5.7 is a cross section of a nano-optic quarter waveplate, which was fabricated by UV-nanolithography [6,32,35,38–40]. It is used in the optical pickup unit within a DVD player. Figure 5.8 shows a performance mapping of the phase retardation distribution on a 4" in-diameter 650 nm quarter waveplate. A very uniform phase retardation of $90° \pm 2°$ across the 4" in-diameter wafer was achieved, along with a high transmittance of >98.8% across the whole wafer. The waveplate also passed 1000 h of environmental test at 85°C and 85 RH.

In Figure 5.9, we show the individual final on-wafer optical performance yields for two batches of a total of 22 4" nano-optic waveplate wafers. On an average, we achieved a final yield of 86%, with a standard deviation of 7%.

5.4.4 LEDs with Enhanced Light Brightness

There is a growing interest in pattern photonic nanostructures (e.g., photonic crystal structure) on top of semiconductor and organic light-emitting diodes to enhance the extraction efficiency of light. Considering the huge total wafer area in various LED manufacturing, NIL was identified as one of the enabling technologies. Luminux is one of the players in this area.

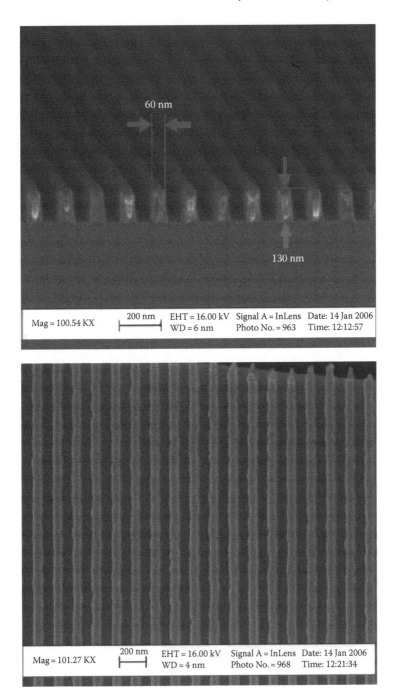

FIGURE 5.6   Four inch diameter visible polarizers made by a UV-NIL.

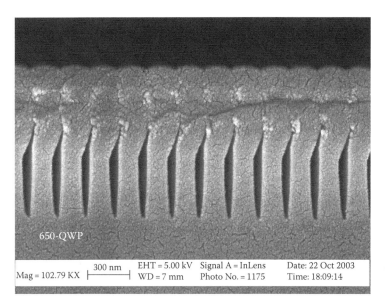

FIGURE 5.7   SEM photography of an optical quarter waveplate for 650 nm.

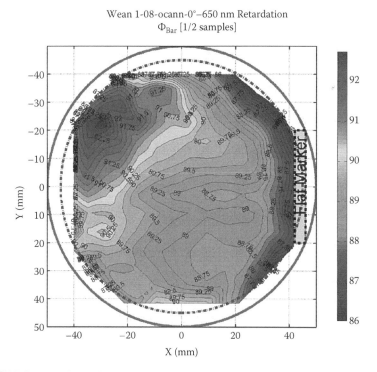

FIGURE 5.8   A wafer performance mapping of a 4" in-diameter nano-optic 650 nm quarter waveplate wafer made by UV-NIL. The map shows the phase retardation distribution uniformity.

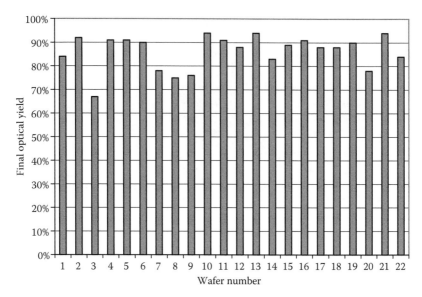

FIGURE 5.9 On-wafer final optical performance yields of two batches of total 22 4″ nano-optic quarter waveplate wafers. The average yield of the 22 wafers is 86%, with a standard deviation of 7%.

5.4.5 MICROOPTICS (MICROLENS ARRAY) AND DIFFRACTIVE OPTICAL ELEMENTS (DOEs)

NIL has a unique advantage in the fabrication of microoptic devices, such as the refractive microlens array as well as various diffractive optical elements (DOEs). Such devices [41–44] typically require complicated curved and/or multilevel patterning with a high resolution. Figure 5.10 shows an imprinted array of microlenses with high-density packing. This is designed for the CMOS/CCD image chip to improve the optical collection efficiency for digital camera applications. The patterning of high packing density lens arrays is difficult to do with conventional lithography technologies. With nanoimprinting, the microlens array can be replicated in a single step. If the polymer can be used as the final lens, no further etching is required. Otherwise, patterned polymer lenses can be used as etching masks to transfer the pattern shapes through deep etching into the material underneath, such as glass. Furthermore, customized aspheric lens shapes can be created. Nanoimprinting can also be done directly into "functional" polymer materials such as the one optimized for a specific optical index of refraction [14].

5.4.6 MULTILAYER INTEGRATED OPTICS

Multilayer and multipixel integrated optical and photonic devices can be made by NIL [28,29,45,46]. Figure 5.11 shows a circular polarizer based on stacking two layers of nanograting-based optical devices, fabricated by NIL.

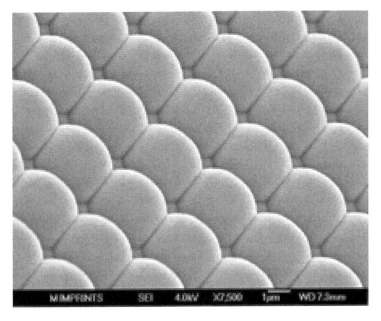

FIGURE 5.10   Microlens array made by NIL.

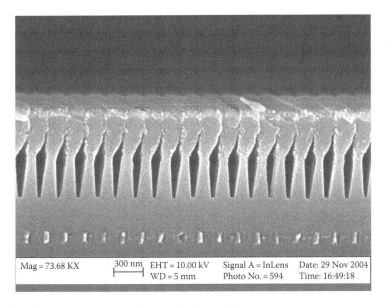

FIGURE 5.11   A two nanolayer circular visible polarizer.

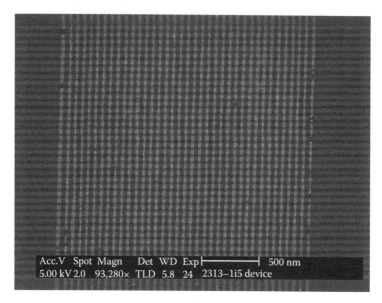

Acc.V Spot Magn   Det WD Exp⊢────────────┤ 500 nm
5.00 kV 2.0   93,280×   TLD 5.8   24   2313–1i5 device

FIGURE 5.12 SEM image of a 34×34 cross-bar memory circuit with 30 nm half-pitch fabricated by NIL. (Courtesy of Dr. Wei Wu of HP Labs.)

5.4.7 MOLECULAR ELECTRONIC MEMORY

HP labs have developed [24] (Figure 5.12) a process to fabricate a crossbar structure using UV-curable NIL with a UV-curable double-layer spin-on resist, metal lift off, and Langmuir–Blodgett film deposition. This process allowed HP to produce 1 kbit cross-bar memory circuits at a 30 nm half-pitch on both top and bottom electrodes.

5.4.8 OPTICAL AND MAGNETIC DATA STORAGE

Data storage is the other wide field NIL is being applied to. With data tracks separated by 108 nm and a minimum pit length of 58 nm, a capacity of 220 GB on a single-layer read-only optical disc is possible [47]. Mass production of such high density discs requires nanoimprint technology. Most recently, Hitachi reported [20] successfully patterning dense concentric data tracks on 2.5" optical disks with thermal nanoimprint technology at the Optics & Photonics 2005 Conference. The concentric data tracks have a width of about 80 nm and a spacing of about 100 nm.

NIL demonstrated fabrication of patterned magnetic media (also called QMD— quantum magnetic disk) a few years ago. Recent media coverage claims that NIL is being developed to fabricate finer nanostructures for magnetic heads and magnetic recording medium, such as patterned magnetic media, by Seagate and Komag. NIL could potentially extend hard drive recording capacities beyond 160 gigabytes per disk, and the storage industry may begin high-volume NIL manufacturing as soon as 2007.

5.4.8.1 Biochips. Bio/Chemical-Sensors, and Microfluidic Devices

NIL offers significant advantages in making nano- and micropatterns for bio- and microfluidic devices and sensors [16,48–50]. One unique capability of NIL is to direct pattern biocompatible materials [49]. Waseda University and a Japanese medical equipment company are developing a cell-sorting device that uses imprinted components to analyze fluids quickly and identify specific target cells.

5.5 SUMMARY

Based on mechanical replication, NIL is an emerging technology that can achieve lithographic resolutions (beyond the limitations set by light diffractions or beam scatterings in conventional lithographic techniques), while promising high-throughput patterning. This chapter reviews the status and some of the recent progress in the commercial applications of this technology.

ACKNOWLEDGMENT

We greatly appreciate contributions from the following colleagues: Lei Chen, Stephen Tai, Xuegong Deng, and Paul Sciortino, Jr. Our appreciation also goes to Greg Blonder, Doug Jamison, Barry Weinbaum, Howard Lee, Hubert Kostal, and Hope Conoscente for their encouragement and support—and to Barbara Shaffer for her time spent on editing the paper.

REFERENCES

1. S. Y. Chou et al., Imprint of sub-25 nm vias and trenches in polymers, *Appl. Phys. Lett.* **67**, 3114 (1995).
2. S. Y. Chou et al., Sub-10 nm imprint lithography and applications, *J. Vac. Sci. Technol. B* **15**, 2897 (1996).
3. J. Haisma et al., Mold-assisted nanolithography: A process for reliable pattern replication, *J. Vac. Sci. Technol. B* **14**, 4129 (1996).
4. M. Colburn et al., Step and flash imprint lithography: A new approach to high resolution patterning, *Proc. SPIE*, **3676**, 379 (1999).
5. M. Bender et al., Fabrication of nanostructures using an UV-based imprint technique, *Microelectron. Eng.* **53**, 236 (2000).
6. J. Wang et al., Wafer based nano-structure manufacturing for integrated nano-optic devices, *J. Lightwave Technol.* **23**, 474 (2005).
7. L. J. Guo, Recent progress in nanoimprint technology and its applications, *J. Phys. D: Appl. Phys.* **37**, R123 (2004).
8. T. S. Sotomayor (ed.), *Alternative Lithography*, Kluwer Academic Publishers, Boston, MA, 2003.
9. A. S. Farber and J. Hilibrand, Method of preparing portions of a semiconductor wafer surface for further processing, United States Patent, US 4,035,226 (1977).
10. L. S. Napoli and J. P. Russell, Process for forming a lithography mask, United States Patent, US 4,731,155 (1988).
11. H. W. Deckman and J. H. Dunsmuir, Procedure for fabrication of microstructures over large areas using physical replication, United States Patent, US 4,512,848 (1985).
12. K. Kazuhiro, M. Yuji, and M. Yoshiro, Method of manufacturing a thin-film pattern on a substrate, United States Patent, US 5,259,926 (1993).

13. J. Wang et al., Fabrication of a new broadband TM-pass waveguide polarizer with a double-layer 190 nm metal gratings using nanoimprint lithography, *J. Vac. Sci. Tech.* **17**, 2957 (1999).
14. J. Wang et al., Direct nanoimprinting of sub-micron organic light-emitting structures, *Appl. Phys. Lett.* **75**, 2767 (1999).
15. M. T. Li et al., Fabrication of circular optical structures with a 20 nm minimum feature size using nanoimprint lithography, *Appl. Phys. Lett.* **76**, 673 (2000).
16. H. Cao et al., Fabrication of 10 nm enclosed nanofluidic channels, *Appl. Phys. Lett.* **81**, 174 (2002).
17. M. T. Li et al., Large area direct nanoimprinting of SiO_2–TiO_2 gel gratings for optical applications, *J. Vac. Sci. Technol. B* **21**, 660 (2003).
18. S. V. Sreenivasan et al., Enhanced nanoimprint process for advanced lithography applications, *Semiconductor Fabtech*, 25th edition, 107 (2005).
19. M. D. Austin et al., 6 nm half-pitch lines and 0.04 μm^2 static random access memory patterns by nanoimprint lithography, *Nanotechnology* **16**, 1058 (2005).
20. T. Ando et al., Development of nanoimprint lithography and applications, *SPIE Optics & Photonics*, San Diego, CA, 2005.
21. H. F. Hess et al., Inspection of templates for imprint lithography, *J. Vac. Sci. Technol. B* **22**, 3300 (2004).
22. W. J. Dauksher et al., Repair of step and repeat imprint lithography templates, *J. Vac. Sci. Technol. B* **22**, 3306 (2004).
23. B. Heidari et al., Combination of electron beam and imprint lithography for realization of pattern media and next generation optical media, *The 49th International Conference on Electron, Ion and Photon Beam Technology and Nanofabrication*, Orlando, FL, 2005.
24. W. Wu et al., One-kilobit cross-bar molecular memory circuits at 30-nm half-pitch fabricated by nanoimprint lithography, *Appl. Phys. A* **80**, 1173 (2005).
25. L. Chen et al., Defect control in nanoimprint lithography, *J. Vac. Sci. Tech.* **B 23**(6), 2933–2938, (2005).
26. W. Zhang and S. Y. Chou, Multilevel nanoimprint lithography with submicron alignment over 4 inch Si wafers, *Appl. Phys. Lett.* **79**, 845 (2001).
27. S. Y. Chou, Fluid pressure imprint lithography, United States Patent Application 20020132482 (2002).
28. J. Wang et al., Subwavelength optical elements (SOEs)—A path to integrate optical components on a chip, *2002 Technical Proceedings, National Fiber Optic Engineers Conference*, Dallas, TX, 2002, p. 1144.
29. J. Wang et al., Design and realization of multi-layer integrated nano-optic devices, *2003 Technical Proceedings, National Fiber Optic Engineers Conference*, Dallas, TX, 2003, p. 785.
30. J. Wang et al., High-performance nanowire-grid polarizers, *Opt. Lett.* **30**, 195 (2005).
31. J. Wang et al., Innovative high-performance nanowire-grid polarizers and integrated isolators, *IEEE J. Select. Top. Quantum Electron.* **11**, 241 (2005).
32. J. Wang et al., Nano-optical devices and integration based on nano-pattern replications and nanolithography, *Proc. SPIE* **5592**, 51 (2005).
33. J. Wang et al., Innovative nano-optical devices, integration and nano-fabrication technologies (invited paper), *Proc. SPIE* **5623**, 259 (2005).
34. J. Wang et al., Free-space nano-optical devices and integration: Design, fabrication, and manufacturing, *Bell Labs Tech. J.* (Nanotechnology issue), **10**(3) (October 2005).
35. K. D. Lee, Visible polarizer by imprint lithography, Nanoimprint and Nanoprint Conference, Vienna, Austria, 2004.
36. J. Wang et al., High-performance large-area ultra-broadband (UV to IR) nanowire-grid polarizers and polarizing beam-splitters, *Proc. SPIE* **5931**, 59310D (2005).

37. J. Wang et al., Monolithically integrated circular polarizers with two-layer nano-gratings fabricated by imprint lithography, *J. Vac. Sci. Tech.* **B 23**(6), 3164–3167 (2005).

38. J. Wang et al., High performance 100 mm in diameter true zero-order waveplates fabricated by imprint lithography, *J. Vac. Sci. Tech.* **B 23**(6), 2950–2953 (2005).

39. X. Deng et al., Achromatic wave plates for optical pickup units fabricated by use of imprint lithography, *Opt. Lett.* **30**(19), 2614–2616 (2005).

40. J. Wang et al., High-performance optical retarders based on all-dielectric immersion nano-gratings, *Opt. Lett.* **32**, 1864 (2005).

41. H. P. Herzig (ed.), *Micro-Optics: Elements, Systems and Applications*, Taylor & Francis, London, 1997.

42. M. T. Gale, Replication technology for micro optics and optical microsystems, *Proc. SPIE* **5177**, 113 (2003).

43. M. Rossi et al., Wafer scale micro-optics replication technology, *SPIE Conference on Lithographic and Micromachining Techniques for Optical Component Fabrication*, San Diego, CA, 2003.

44. M. Rossi and I. Kallioniemi, Micro-optical modules fabricated by high-precision replication processes, OSA topical meeting "Diffractive optics and micro-optics", Tucson, AZ, paper DTuC1, TOPS 75 (2002).

45. H. Kostal and J. Wang, Nano-optic devices enable integrated fabrication, *Laser Focus World*, **40**(6), 155–159 (June 2004).

46. J. Wang et al., Integrated nano-optic devices based on immersion nano-gratings made by imprint lithography and nano-trench-filling technology, *Proc. SPIE* **5931**, 59310C.1–59310C.12 (2005).

47. T. D. Milster, Horizons for optical data storage, *Opt. Photon. News*, **16**, 28–33 (March 2005).

48. L. J. Guo et al., Fabrication of size-controllable nanofluidic channels by nanoimprinting and its application for DNA stretching, *Nano Lett.*, **4**(1), 69–73 (2004) (communication).

49. J. D. Hoff et al., Nanoscale protein patterning by imprint lithography, *Nano Lett.* **4**, 853–857 (2004) (communication).

50. J. Wang et al., Resonant grating filters as refractive index sensors for chemical and biological detection, *J. Vac. Sci. Tech.*, **B 23**(6), 3006–3010 (2005).

6 Design and Fabrication of Planar Photonic Crystals

Dennis W. Prather, Ahmed Sharkawy,
Shouyuan Shi, Janusz Murakowski,
Garrett Schneider, and Caihua Chen

CONTENTS

6.1 INTRODUCTION

The fundamental design architecture for electrical integrated circuits consists of devices arranged on a planar semiconductor substrate that are connected by metallization traces for in-plane communication and vias for interconnecting multiple planes. Along these lines, planar chip architectures have become the universal embodiment for integrated circuit design and, as a result, have given rise to a standardization in microfabrication techniques. While there is currently a strong research effort in the semiconductor community to find alternatives to the obstacles facing modern-day integrated circuits, semiconductor-based optical integrated circuits (OICs) provide a paradigm shift for next-generation integrated circuits. Replacing metallic wires with optical waveguides has revolutionized the telecommunication industry through the introduction of fiber optic cables and has been long sought after as an alternative for electrical integrated circuits.

In general, dielectric waveguides in planar OICs commonly guide or manipulate light via total internal reflection (TIR). For example, ridge and rib waveguides guide light by means of TIR in both the lateral and vertical dimensions. However, as with electrical integrated circuits, a similar limitation arises when one begins to increase the waveguide integration density. The fact is, as optical waveguides get smaller and closer to each other, i.e., the integration density increases, the amount of cross talk between these waveguides increases due to the fact that an increasing amount of energy is transferred into the evanescent tails of the propagating waveguide mode. While this is useful for creating integrated optical devices such as directional couplers and fiber-waveguide input couplers, this poses a severe limitation when attempting to reduce the overall size of the chip. However, this limitation can be overcome by employing photonic crystal (PhC) waveguides due to the difference between the guiding mechanisms in conventional and PhC waveguides. The difference lies in the fact that two-dimensional PhC waveguides guide light in the plane, i.e., laterally, by distributed Bragg reflection (DBR) due to the periodic nature of the PhC [1].

While a complete OIC consists of imbedded sources, routing and switching devices, and detectors, this chapter focuses on the realization of nanoscale PhC-based guiding, routing, and switching devices for on-chip communication.

6.2 PHC FUNDAMENTALS

In the past decade, there has been a growing interest in the realization of PhCs as optical components and circuits, which is mainly due to their unique ability to control the propagation of light. In 1987, Yablonovitch [2] and John [3], proposed that a periodic dielectric structure can possess the property of a photonic bandgap (PBG) for certain regions in the electromagnetic spectrum, in much the same way an electronic bandgap exists in semiconductor materials. Due to their analogy with electronic semiconductor materials, these structures were named "PhCs." Examples of a one-, two-, and three-dimensional PhC structures are shown in Figure 6.1.

While one-dimensional thin-film stacks have been known for over a century, their generalization to higher dimensions was not proposed until the 1970s by Bykov [4,5] as a possible means for inhibiting spontaneous emission. One-dimensional PhCs are more commonly referred to as a type of distributed Bragg reflector, commonly found in quantum well devices and antireflective coatings, and thereby only have an effect on the direction of light propagation normal to the surface of the structure shown in Figure 6.1a. Three-dimensional PhCs, as shown in Figure 6.1c, provide an ideal platform for guiding and manipulating light in all dimensions [6]; however, their

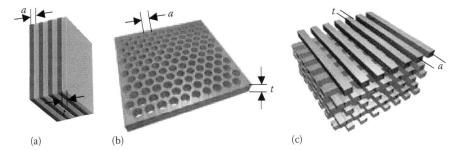

FIGURE 6.1 A schematic of a (a) one-, (b) two-, and (c) three-dimensional periodic structure, where a is the lattice constant and t is the thickness of the layer. When properly designed, such structures exhibit the property of a photonic bandgap (PBG) for certain band of frequencies and lattice orientation.

usefulness is limited by the difficulty of fabricating and introducing defects into such a structure. To this end, two-dimensional PhCs have been explored extensively, both theoretically [7–11] and experimentally [12–15], primarily due to the ease in which they are fabricated for optical frequencies. This ease exists because introducing and precisely controlling defects, such as waveguides or cavities, are primarily defined through standard lithographic techniques. Two-dimensional PhCs are generally realized either by a periodic array of dielectric rods in air, or by perforating a dielectric slab with air holes of any shape and/or geometry. Such structures can be optimized either structurally or through material modifications to manipulate the size and location of the PBG as well as the inherent unique dispersive characteristics. Two-dimensional PhCs impose periodicity within the plane (lateral dimension) while the third dimension is either infinitely long such as in a PhC fiber or is finite in height as with a PhC slab shown in Figure 6.1b.

6.2.1 CRYSTALLINE TERMINOLOGY

The etymology of the term "PhC" stems from the analogy with the electrical characteristics of crystalline materials. A crystal is defined as "a body that is formed by the solidification of a chemical element, a compound, or a mixture and has a regularly repeating internal arrangement of its atoms and often external plane faces." From an electron's perspective, this periodic arrangement of atoms or molecules represents a periodic potential in which the macroscopic or bulk conduction properties are determined by the atomic composition of the crystal. The Bragg-like diffraction imposed by the periodic arrangement of atoms introduces gaps in the energy band such that electrons are forbidden to propagate at these energies in certain directions. Furthermore, from quantum mechanics, we know that the wave-like propagation of a particle, such as an electron, in a crystal must obey the Schrödinger equation.

Similarly, a PhC is a periodic arrangement of dielectric materials, arranged in such a way to manipulate the macroscopic electromagnetic properties of the material.

The term most commonly regarded as the fundamental macroscopic property of the PhC is the PBG, which is analogous to the electronic bandgap found between the valence and conduction bands in a semiconductor. The PBG describes the energy (frequency) band for photons that are not permitted to propagate inside the crystal in certain directions due to the Bragg-like diffraction of the photon. To draw the analogy even further, as the electron obeys the Schrödinger equation in an electronic crystal, the propagation of a photon through a PhC must obey Maxwell's equations. To this end, the fundamental dependent variables of a PhC are the materials, more specifically their inherent electromagnetic properties, and their respective structural arrangement.

If the materials are chosen with a large difference in relative permittivities, then the phenomena commonly observed due to the coherent scattering of electrons in semiconductor crystals can also be observed from the coherent scattering of photons in PhCs. The types of materials most commonly employed are dielectrics, namely standard thin-film dielectrics in one dimension, semiconductors in two dimensions, and polymers in three dimensions, while metals are sometimes used. As shown in Figure 6.1a, in one dimension, the PhC consists of a periodic arrangement of material planes, in which the design is limited to the thickness and choice of materials for each of these planes. However, the degrees of freedom in two-dimensional PhCs allow for the formation of more complex structures that are laterally periodic in two dimensions and vertically homogeneous in the third as shown in Figure 6.1b. In two dimensions, the number of dependent parameters increases to include the crystalline lattice orientation, shape, and size of the perturbation, i.e., lattice site, spacing between lattice sites, i.e., the lattice constant, and, of course, the electromagnetic properties of all constituent materials involved. In three dimensions, the types of structures are periodic along three axes and are much more complex due to the added dimension. In this section, the various dependent parameters of a PhC are discussed to provide a foundation for how they individually and collectively affect the macroscopic properties of the PhC. However, before examining the effect of these structural modifications on the PBG, the associated crystalline terminology and the related computational methods will be discussed.

The development of the first theories of PhC analysis was largely performed by solid-state physicists [16]. For this reason, many of the descriptive terms for PhCs are analogous to similar references in solid-state physics such as the reciprocal lattice, Brillouin zone, and dispersion diagrams. Therefore, these terms will be briefly discussed to provide a foundational understanding.

Figure 6.2a shows a two-dimensional square PhC lattice in which a is the lattice constant and $a\hat{x}$ and $a\hat{y}$ are the primitive lattice vectors. The reciprocal space (sometimes called wave vector or k-space) is shown in Figure 6.2b where the primitive reciprocal lattice vectors are defined as $(2\pi a)\hat{x}$ and $(2\pi a)\hat{y}$. The k-space is used for analyzing light propagation in PhCs since it is the most appropriate space for plotting wave vectors. By representing the wave vectors in k-space, one can more visibly understand the means by which the dispersion relation is calculated. However, in order to clearly understand the dispersion relation, we separate the k-space into Brillouin zones. The boundaries of the Brillouin zones in k-space

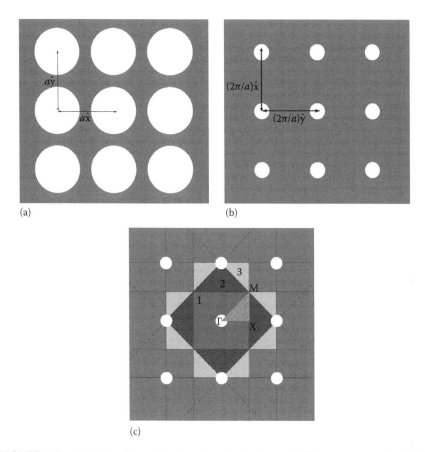

FIGURE 6.2 (a) A two-dimensional rectangular lattice and (b) its representation in the reciprocal space. (c) Representation of the first (1), second (2), and third (3) Brillouin zones.

are determined by the k-values for which Bragg diffraction occurs. Therefore, the separation of k-space into Brillouin zones is useful because we find that energy gaps (PBGs) appear when the Bragg condition is satisfied. Specifically, the boundaries of the Brillouin zones are the perpendicular bisectors of the reciprocal lattice vectors for the lattice points shown in Figure 6.2b. By doing so, one can easily outline the Brillouin zones in k-space as shown in Figure 6.2c. While the k-space can be split into multiple Brillouin zones such as the first, second, and third zones shown as 1, 2, and 3 in Figure 6.2c, respectively, due to the rotational symmetry in k-space, only a part of the first Brillouin zone need be considered. The irreducible Brillouin zone is defined as the triangular region where the wave vectors at the corners of the triangle are given by $\Gamma = 0$, $X = \pi/a\hat{x}$, and $M = \pi/a(\hat{x} + \hat{y})$ [11,17]. By exploiting the rotational symmetry, we can significantly reduce computational costs for calculating the dispersion relation.

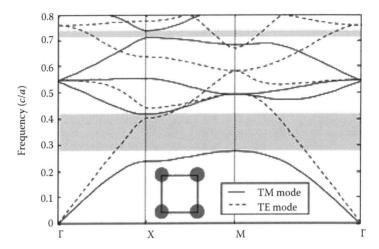

FIGURE 6.3   The dispersion diagram of a rectangular lattice in which the solid lines represent TE-polarization and the dashed lines depict TM-polarization. The two grey bars depict the PBGs for TE-polarization.

In an electrical crystal, the dispersion relation provides a mapping of the energy of an electron as a function of the wave vector of the electron. Analogously, in a PhC, the dispersion relation represents the energy of the photon (frequency) as a function of the photon's wave vector. The solution of an eigenvalue problem, governed by the periodic dielectric distribution, at points on the boundary the irreducible Brillouin zone results in the dispersion diagram as shown in Figure 6.3. Such a dispersion diagram can be generated by solving this eigenvalue problem with various computational methods.

6.2.2 LATTICE TYPES

A fundamental knowledge of crystallography is an indispensable tool for solid-state physicists, material scientists, and chemists. In standard three-dimensional crystals, there are numerous crystal lattice structures ranging from the elementary, such as the various cubic and closed-packed models, to intricate structures such as alloys and organic compounds. While there are a large number of complex crystal lattice structures in three dimensions, limiting the dimensionality naturally reduces the possibilities. Although many lattice structures exist in two dimensions, for a two-dimensional PhC, the two most common lattice structures are the triangular (hexagonal) and square as shown in Figure 6.4a and b, respectively. These structures are most common for two reasons: (1) they both can generate large PBGs for certain frequencies and polarizations and (2) defects can be easily introduced into these simple crystalline structures to create optical devices. In order for the crystalline lattice to possess a PBG, there must be a large contrast between the background and lattice materials, i.e., the black and white regions in Figure 6.4, respectively. While many initial works in the field of PhCs concentrated on the simulation and analysis of PhCs consisting of

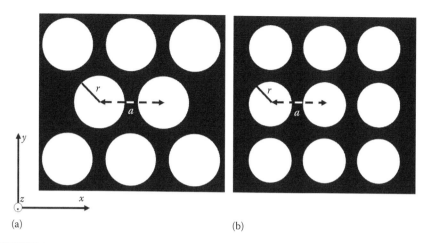

(a)   (b)

FIGURE 6.4 Common lattice structures for two-dimensional PhCs. (a) A triangular lattice, where r is the radius of the perturbation and a is the lattice constant. (b) A square PhC lattice.

dielectric rods in air [18,19], as shown in Figure 6.5a, a fundamental reason swayed the focus to the opposite case in which the perturbations had a lower relative permittivity with respect to the background. This fundamental reason is that when attempting to realize planar PhCs, such that the thickness of the plane is finite in extent, the light must be sufficiently confined to this plane. Therefore, in order to achieve this, researchers began studying PhC structures defined in high dielectric slabs [20,21], such as the PhC structure shown in Figure 6.5b. Therefore, when light is laterally localized in the material comprising the defect region, i.e., black in Figure 6.4, this material must have a larger relative permittivity than the cladding material in the z-dimension, resulting in the slab structure shown in Figure 6.5b.

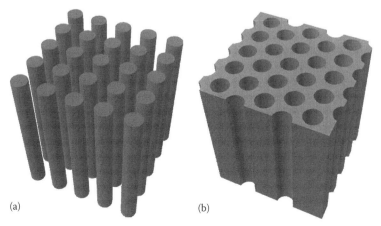

(a)   (b)

FIGURE 6.5 Common embodiments of a PhC in two dimensions. (a) Dielectric rods in air (lower relative permittivity) background. (b) Air holes in a material with larger permittivity.

6.2.3 Computational Methods

The ability to extract the spatial and temporal properties of PhC structures was a crucial step for any future progress in the development of functional PhC-based devices and applications. Hence, the initial challenge for researchers within the community to overcome was developing accurate yet fast modeling and simulation tools that were capable of analyzing various globally and locally periodic structures with complex material properties.

During the early 1990s, most of the research efforts had been put toward developing efficient and accurate algorithms to carry out the calculations of the band structures. Among these algorithms, there are several most popular methods, such as plane wave expansion method (PWEM) [22–24], finite-difference time-domain (FDTD) method [25,26], transfer matrix method [27], and finite element method (FEM) [28,29]. Beside these techniques, several interesting approaches have been developed [30–32].

The PWEM represents the most simple and straightforward manner to represent the periodic fields using common Fourier expansion in terms of harmonic functions defined by the reciprocal lattice vectors. This simplicity, together with the development of powerful numerical procedures, has made the plane wave method (PWM) the most widely used tool for finding Bloch modes and eigenfrequencies of infinite periodic system of scattering objects. An application of the Fourier expansion turns Maxwell's equation into an eigenvalue problem. A most widely used version of the method is applying the preconditioned conjugate gradient minimization of the Rayleigh quotient for finding eigenstates and frequencies [23]. The minimization of the Rayleigh quotient allows us to handle thousands of plane waves. However, the PWEM was limited to simulating infinitely periodic structures, which was constrained by multiple symmetries and assumed the structure to be lossless. It also assumed the structure to be perfectly periodic and, hence, fabrication tolerances, which highly modulate the spatial and temporal response or a periodic structure, could not be simulated in a PWM platform. In addition, it was not capable of calculating the transmission and/or reflection spectra. Nevertheless PWEM remains to be a useful platform to quickly determine whether or not a conventional lossless periodic structure may or may not have a bandgap for a specific polarization, and remains to be the platform for extracting the highly complex dispersive properties of such periodic structures.

The FDTD method [33] is widely used to calculate transmission and reflection spectra for general computational electromagnetic problems, and it is considered to be one of the most applicable for the PhCs. It is universal, robust, methodologically simple and descriptive. The wave propagating through the PhC structure is found by direct integration in the time domain of Maxwell's equations in a discretized difference form. Discretization in both time and space is done on the staggered grid. In addition to discretization, the proper boundary conditions, i.e., absorbing and periodic boundary conditions, can be applied. If one defines the input signal as continuous wave (CW) or pulse, the excitation can be propagated through the structure by time stepping through the entire grid repeatedly. Several algorithms had been developed to calculate the band structures. However, the basic FDTD implementation on a single computer was an extremely slow solution, a particularly problematic issue that grows exponentially with the complexity of the devices under investigation. Various approaches have been taken to tackle this problem. The initial solution was to parallelize the FDTD algorithm over a Beowulf cluster of tens or hundreds of personal computer (PC) nodes; this, however, provided only a short-term resolution to the ongoing issue of computational time. The mean time to failure of the number of nodes constituting the cluster grew exponentially as did the maintenance cost and the physical space necessary to host such clusters. Recently, a different approach to solve the same problem was taken. This approach relies on implementing the FDTD algorithm over a hardware accelerator-based workstation [34–37], where a dedicated field-programmable gate array (FPGA)-based chip is programmed to execute the FDTD algorithm at computational speeds equivalent to 150 PC node cluster. Such an approach is expected to become the standard for modeling and simulation of highly complex PhC-based structures and devices.

In the similar manner to the FDTD method, the transfer matrix method is implemented by discretizing Maxwell's equations. However, the initial excitation is supposed to be monochromatic, and the basic structure unit is the layer of grid cells. The structure under consideration is divided into the set of layers with the same number of grid nodes in each layer. Using the discretized Maxwell's equations, the field $\mathbf{E}_i$ in the nodes of one layer may be connected to the field $\mathbf{E}_{i+1}$ in the nodes of the neighboring layers via the transfer matrix $\mathbf{E}_{i+1} = \mathbf{T}_i \mathbf{E}_i$. Thus, by integrating all layers, the output field is connected to the input field by the transfer matrix, which is a product of individual layer-to-layer transfer matrices. Also similar to the FDTD method, proper boundary conditions should be used. The transfer matrix method is less universal due to numerical instability during the integration; however, it is more computationally effective than the FDTD method since it is an on-shell method. The method can be used to model infinite periodic structures, and to find the eigenmodes and eigenfrequencies, as well as transmission properties.

FEM is a frequency domain method used to solve Maxwell's equation. In fact, it is based on a variational principle, which is the same as the PWM. Instead of using the plane waves as expansion basis, which is defined in entire unit cell, FEM use subdomain basis to discretize the computational unit cell. FEM takes into account discontinuities in the dielectric function to overcome the slow convergence of PWEM. To solve the matrix eigenvalue problem, a preconditioned subspace iteration algorithm may be applied to find a relative small number, say p, of the interest smallest eigenvalues in large-dimensional symmetric positive defined matrix problems.

One very promising class of PhC structures is the PhC slab, which has two-dimensional (in-plane) periodicity; the height is finite and comparable to the wavelength of light. An example is shown in Figure 6.6a. The PhCs slab is relatively easier to fabricate than three-dimensional PhC structures and more attractive for chip-level integration of different optical devices [38–40]. Being finite in height requires another mechanism for light confinement in the third dimension, namely TIR. It is then the combination of these two phenomena the in-plane light confinement through multiple Bragg reflections or some particular dispersion properties and the finite height of the structure that gives PhC slabs their genuine advantage for usage in planar photonic integrated circuits.

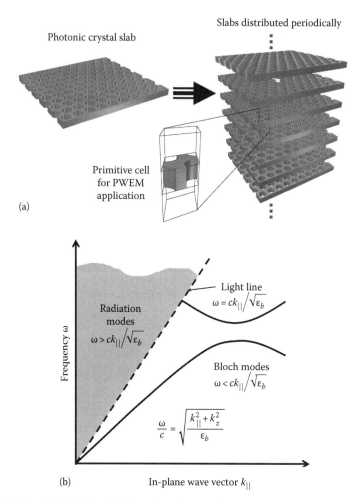

FIGURE 6.6 (a) PhC slab, (b) dispersion diagram for in-plane wave vector of a periodic structure overlapped with light line.

The boundary between the guided and radiation modes is described as the light line. The radiation modes are the states extended infinitely in the clad region outside the slab, and the guided modes are those localized to the plane of the slab, as shown in Figure 6.6b. States that lie below the light line in the band diagram cannot couple with modes in the bulk background. Thus, the discrete bands below the light line are confined. Mathematically, we express the wave vector k as $\mathbf{k} = \mathbf{k}_\parallel + \mathbf{k}_z$, where $\mathbf{k}_\parallel$ is the in-plane wave vector, and $\mathbf{k}_z$ is the out-plane wave vector. If the guided modes have the imaginary k_z component, then their modes decay in the cladding. However, if the radiation modes have a real k_z component, then they will leak to the cladding.

PhC slab structures have many potential applications and most of them rely on their corresponding band structures. To employ the PWM for the band structure calculations of PhC slab, which has only two-dimensional periodicity, a third dimensional periodicity was imposed by introducing a periodic sequence of slabs separated by a sufficient amount of background region to ensure electromagnetic isolation, namely the supercell technique. The guided modes are localized within the slab so that the additional periodicity of periodic slabs with large separation will not affect their eigenfrequencies. However, for the radiation (leaky) modes, which lies above the light line, this technique is no longer appropriate due to the artificial periodicity in the out-plane direction.

Determining the leaky modes above the light line requires the application of a perfectly matched layer (PML) in the z-direction to absorb the radiations from the slab. The PML absorbing boundary condition [41], which was firstly introduced by Berenger into FDTD, has been proven as the most powerful way to absorb waves for any frequency and angle of incidence. The anisotropic material-based formulation offers special advantage in that it does not require modification of Maxwell's equations [42]. With the PMLs, the artificial periodicity in the z-direction can be introduced without affecting the problem. Combining these techniques, PWM was used to cast Maxwell's equations into a generalized complex eigenvalue problem [43].

The introduction of the PMLs will sufficiently suppress spurious modes. However, the undesired so-called PML modes will be generated due to the periodic boundary conditions applied along the z-direction. Therefore, an additional tool is required to distinguish the guide modes, leaky modes, and PML modes. Two concepts can be used to distinguish those modes: one is based on the Q factor of complex resonance mode, and the other is based on the fact that the guided modes are characterized by a high power concentration in the PhC slab.

Consider a square lattice with air holes embedded in slab with a high dielectric constant of 12.25 and a thickness of $0.6a$, as shown in Figure 6.7a. The air holes are of a circular cross-section with radius $r = 0.3a$. As shown in Figure 6.7b, there is a good agreement between the modified PWM with PML method and a three-dimensional FDTD method.

Over the past 15 years, numerous modeling and simulation tools to design and analyze complex PhC structures were introduced in the commercial market. These tools reflected ongoing research activities within academic and industrial institutions and were the initial building block for the true development of functional PhC-based applications. Each tool has its advantages and disadvantageous for certain research scenarios. With a wide range of available tools in hand, the community next moved toward realization of various devices and applications synthetically implemented using such tools.

Since PhCs can be designed for operation in semiconductor materials, it would be prudent to leverage the existing microfabrication techniques for the realization of planar PhC devices and circuits. The ability to rapidly fabricate and characterize planar PhCs makes them a desirable platform for analyzing the unique properties inherent to PhCs. To this end, this chapter is limited primarily to the discussion of planar PhC devices and circuits. In the case of chip-scale optical networks, an optical infrastructure must be realized on a scale that is commensurate with the submicron dimensions of chip-scale components. For this reason, PhC technology offers the promise of realizing such chip-scale photonic networks. However, for chip-scale

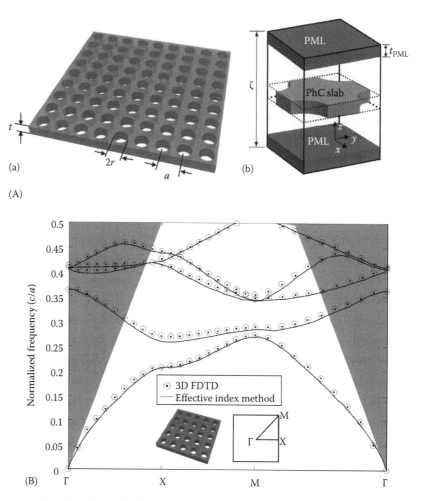

FIGURE 6.7   (A) Unit cell for the band structure calculations. (B) Comparison of dispersion diagram between the FDTD and the presented methods for even mode.

photonic networks to be realized, several technological barriers must be overcome. Such barriers include repeatable and reliable design tools, high-fidelity fabrication processes, high-efficiency input and output coupling structures, and manufacturable integration processes. While each of these issues is being addressed in the research community, in the next section, we discuss the realization of planar PhCs from design and conceptualization through fabrication processes developed to date.

6.3   PROTOTYPING PLANAR PHCS

Performing lithography is the first step toward prototyping PhCs. The dimensions of PhC devices are determined by the wavelength of light in the medium for which they are designed $\lambda = \lambda_0/n$, where λ_0 is the free-space wavelength and n is the index of refraction of the medium. An example is silicon which has a refractive index $n_{Si} \sim 3.5$. Silicon devices designed to operate at free-space wavelength of 1550 nm have features of approximately 450 nm, and those designed to operate at 1330 nm have dimensions on the order of 350 nm. These dimensions are difficult to achieve using most commonly available standard photolithography equipment in academic research labs.

In addition, unlike electronic devices, critical dimension (CD) control and smooth profiles are extremely critical in the operation of photonic devices, as light interaction with any foreign matter or unwanted defects leads to scattering losses, inhibiting the device's performance. Although state-of-the art lithography equipment is capable of achieving dimensions of ~350 nm, sufficient for typical integrated optical devices, nanophotonic devices like ring resonators and PhCs require CD values of less than 100 nm (membranes between the holes of a PhC lattice). Only recently, semiconductor industry leaders have been able to faithfully fabricate at the 90 nm technology node used in Pentium 4 chips, which explains the reason for the slow progress in PhC product development at optical frequencies. The reticle costs alone for such a technology node exceed a million dollars, which places it beyond the reach of most academic research labs [44]. On the other hand, electron-beam lithography (EBL), while unsuitable for mass production, is an effective method for research and development, and prototyping of nanoscale devices.

6.3.1 ELECTRON-BEAM LITHOGRAPHY PROCESS

A Raith-50 EBL system was used to pattern the PhC lattice in polymethyl methacrylate (PMMA). PMMA serves as both positive contrast e-beam resist and as an etch mask, without requiring intermediate transfer steps, which improves the overall process robustness. Minimizing the number of fabrication steps in the development of nanoscale devices is extremely critical as every additional step increases the complexity and reduces the yield and performance of the device. This led to the development of the fabrication methodology (pattern transfer into the substrate) presented in details below, using the e-beam resist as the etch mask.

Figure 6.8 depicts the entire process flow for the fabrication of two-dimensional PhCs on silicon-on-insulator (SOI) substrates.

The electron dose must be precisely controlled while exposing PMMA in order to achieve the desired diameter of the holes in the PhC lattice. One effect that complicates the exposure definition is that the electrons scatter as soon as they penetrate the PMMA, resulting in a wider exposed area deep in the material as compared to the area close to the surface. This phenomenon of electron scattering is known as the proximity effect and is usually undesirable as it limits the resolution of lithography. However when exposing hexagons, this scattering serves to generate circular structures.

While circles are obviously desired for PhCs due to their axial symmetry, hexagons were used in the design because the overall time taken to expose the entire structure, i.e., the write time, increases considerably with the increasing number of vertices contained in the entire CAD layout. Because circles can contain nearly an

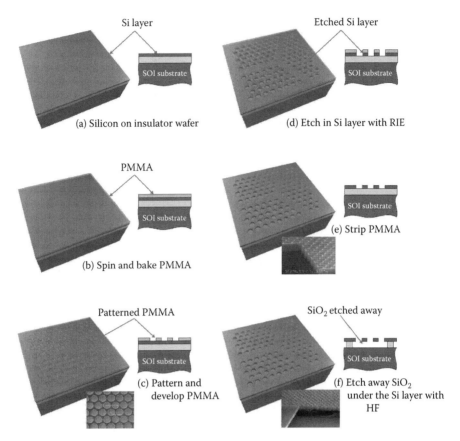

Si layer

SOI substrate

(a) Silicon on insulator wafer

PMMA

SOI substrate

(b) Spin and bake PMMA

Patterned PMMA

SOI substrate

(c) Pattern and
develop PMMA

Etched Si layer

SOI substrate

(d) Etch in Si layer with RIE

SOI substrate

(e) Strip PMMA

SiO_2 etched away

SOI substrate

(f) Etch away SiO_2
under the Si layer with
HF

FIGURE 6.8 Process flow for fabricating PhCs in an SOI wafer. (a) Unprocessed SOI wafer. (b) Spin a 200 nm-thick layer of PMMA (e-beam sensitive resist). (c) Pattern PMMA with e-beam and develop exposed regions. (d) Etch Si device layer with anisotropic reactive-ion etch. (e) Remove PMMA. (f) Underetch SiO_2 layer in hydrofluoric acid to create a suspended silicon PhC membrane.

order of magnitude more vertices than a hexagon, this difference would generate undesirably long write times. Moreover, since the beam current fluctuates over time, a shorter write time was desired to achieve a uniform dose profile over the entire structure. For these experiments, submicron high-fill-factor ($r/a > 0.4$) PhC lattice structures were obtained in PMMA using a 100 pA beam current and 20 kV accelerating voltage.

After exposure, PMMA was developed in a 1:3 solution of methyl isobutyl ketone (MIBK) in isopropyl alcohol for 30 s. The developer dissolved the regions exposed to the electron beam, as depicted in Figure 6.8c. A scanning electron micrograph of a triangular PhC lattice patterned in PMMA is shown in Figure 6.9.

Once a high-resolution lithography process has been developed, a process to transfer the exposed pattern into the underlying substrate, i.e., an etch recipe, must be developed. As can be observed from Figure 6.9, the minimum spacing between

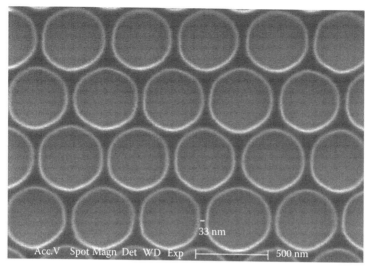

FIGURE 6.9 Scanning electron micrograph of a high-fill-factor ($r/a > 0.4$) triangular PhC lattice exposed in PMMA. The diameter of the holes is approximately 460 nm and the lattice constant, $a = 510$ nm.

the holes of the PhC lattice is on the order of 35 nm, and to fabricate these devices in the 450 nm SOI device layer, requires an aspect ratio (d/w) > 10, where w is the width and d is the depth, which is extremely difficult to achieve using conventional etch methods. In the following section, we briefly discuss conventional etching methods and the need to develop an advanced etch process for the fabrication of PhCs.

6.3.2 Conventional Silicon Etching

Etching is the process of removing unwanted material from the wafer surface by using either chemical or physical means or by the combination of both to accurately reproduce the designed features on the mask into the underlying substrate [45]. The geometries of the etched features lie along a continuum between fully "isotropic" (equal etch rates in all directions) to "anisotropic" (etching along a single direction, with results typically exhibiting perfectly flat surfaces and well-defined, sharp angles). Etching can be classified as either wet or dry etching.

6.3.2.1 Wet Silicon Etching

Among wet etchants, the most common wet silicon etch is "HNA" (mixture of hydrofluoric acid, nitric acid, and acetic acid) [46]. Apart from being isotropic in nature, HNA suffers drastically from doping effects. The etch rate is slowed down by a factor of ~150 in regions of light doping (<10^{17} cm^{-3} n- or p-type) as compared to more heavily doped regions. Furthermore, isotropic wet etching is controlled by diffusion and convection in the etchant and therefore stirring has a considerable effect on the etch profile [47].

Anisotropic (or "orientation-dependent") etchants etch much faster in some crystalline planes than others. Hydroxides of alkali metals (i.e., KOH, NaOH, CsOH, RbOH, etc.) are excellent examples of anisotropic etchants [45]. Some disadvantages of these etchants include crystal-orientation-dependent etching and their dependency on chemical composition, concentration, and temperature.

While they enable "mirror-like" finishes, anisotropic wet etchants generally tend to be extremely beneficial in the development of sloped profiles like V-grooves or prisms. Perfect vertical sidewalls with anisotropic wet etching is possible, however they either necessitate

a. The use of specialized wafers, namely (111) or off-axis cut wafers which are more expensive than the common (100) wafer which are used for active processing of MOS devices
b. A considerable amount of effort in the design of sophisticated mask set patterns to counter orientation-dependent etching [47–49]

Ethylene diamine pyrochatechol (EDP) is another example of an anisotropic and dopant-modulated wet etchant [50]. However, EDP mixtures are potentially carcinogenic and extremely corrosive. Hence their usage in mainstream IC fabrication clean rooms is restricted. In contrast to wet etching, dry etching is the preferred choice in fabrication of microsystems as it offers the ability to realize vertical sidewalls and is largely independent of the crystal orientation or doping.

6.3.2.2 Dry Silicon Etching

The most common form of "dry" etching of silicon is plasma etching, also commonly known as reactive-ion etching (RIE). The fundamental processes involved in the creation of plasma include electronic excitation, dissociation, and ionization [51]. In the electronic excitation process, an electron from an atom is pushed into an orbit at a greater distance from the atom center. The moment the electron falls back to a lower energy level, it emits a photon which gives the plasma a characteristic glow (C_4F_8 plasma is purple in color; O_2 plasma is grayish-white). In the dissociation process, a molecule is broken into smaller species like atoms and radicals which are quite reactive and are responsible for many chemical reactions on the surface of the wafer. In the ionization process, an electron is broken out of its shell and leaves behind an ion which increases the number of charged particles. These ions (responsible for the bombardment of surfaces), along with the radicals that remove surface atoms by chemical reactions, are the most important particles governing plasma etching. Hence to create plasma, a flux of highly energetic particles is needed to be able to ionize the gas. This can be achieved by various means such as photons in laser fusion, chemical energy from exothermic reactions in flames, and electric fields as in gas discharge tubes such as a neon lamp. Among these, the latter method is the most effective in the creation of plasmas due to the fact that the extra electrons formed by ionization will also convert electrical potential energy into kinetic energy, producing even more electrons via collisions. Hence, it is widely used in semiconductor plasma equipment.

The moment when the strength of the electric field is high enough to increase the number of electrons rapidly is called breakdown, and at this moment the gas flashes and becomes plasma. This breakdown voltage is related to the pressure and electrode spacing and is expressed with the help of the Paschen curve [45]. Radio-frequency plasma increases the lifetime of the electrons in the reaction chamber as compared to DC making ionization possible at lower pressures, which is desirable for anisotropic etching. Figure 6.10 depicts typical processes taking place in a reactive-ion etcher.

Plasma processing has been the subject of research for several decades and has attracted investigation from several disciplines of scientific communities due to its profound impact [52]. Physicists describe plasma processes by looking at individual molecules (atomic level) and utilizing concepts like the kinetic theory of gases and statistical mechanics. Chemists focus on thermodynamics and use concepts like enthalpy and entropy to describe plasma. Finally, engineers utilize electrical equivalent circuits to describe ion-enhanced mechanisms and generation of self-bias voltage that govern ion-based etching to describe plasma phenomena [53,54]. From Figure 6.10, one can observe a plethora of complex chemical reactions taking place inside an etch chamber. To evaluate these reactions, researchers use techniques like optical emission spectroscopy [55], and plasma absorption probe [56] to measure the gas-phase physical parameters [57]. To measure dissociated chemical species (radicals), some of the techniques utilized are infrared laser absorption spectroscopy (IR-LAS) [58], laser-induced fluorescence spectroscopy (LIF) [59], quadruple mass spectrometry (QMS) [60,61], and actinometry [62]. Further, ion flux toward the wafer is analyzed with an ion flux energy analyzer (IFEA) [63,64]. Due to the high complexity of the involved processes it is extremely difficult to

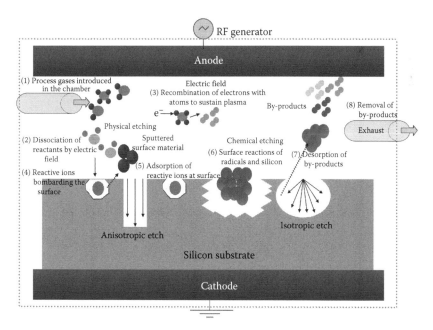

FIGURE 6.10 Typical processes in a reactive-ion etch chamber. (Adapted from Madou, M.J., *Fundamentals of Microfabrication: The Science of Miniaturization*, 2nd edn., CRC Press, Boca Raton, FL, 2001.)

accurately model plasma-etching characteristics as it requires vast infrastructure and a strong interdisciplinary effort. Therefore, the widely adopted methodology in process development (specific gas mixtures or "recipes") is based on empirical studies: trial and error. Here we focused on plasma processing using a process engineering approach by investigating gas chemistries that are compatible with semiconductor industry and scrutinized their processing effects on silicon nano-photonic devices.

6.3.2.3   Effects of Etching on PhC Devices

Several researchers have investigated the effects of etching on PhC properties [65–67]. In particular Noda and coworkers [65] investigated the effects of truncated-cone air holes (i.e., failure to realize vertical sidewalls) in a two-dimensional PhC slab structure. Their results showed that truncated-cone air holes cause large propagation losses that are approximately proportional to the square of the taper angle. Air holes with 5° taper result in a propagation loss of about 77.7 dB/mm, while those with only 1° taper causes a propagation loss of 0.5 dB/mm. For air holes with no taper, light is confined within the waveguide and only TE-like waveguide modes are excited. On the other hand, for tapered air holes, TE-like waveguide modes couple to TM-like slab modes leading to propagation losses. Summarizing the findings of the above references, the finite height of air holes, sidewall roughness, inclination, and nonuniformity of the hole sizes, affect the properties of PhCs significantly by contributing to the propagation loss in line-defect waveguides.

Furthermore, high-aspect-ratio (HAR) anisotropic etching is extremely important in the realization of integrated optical devices using planar PhCs for another reason. Typically, such PhCs consist of arrays of high-fill-factor lattices, where r/a commonly exceeds 0.4. Thus, for devices operating at telecommunication wavelengths, etch processes must be capable of forming sub-100 nm vertical membranes with HAR.

Due to the limitations of conventional etching techniques, there is a strong need to develop a deep reactive-ion etching (DRIE) process for the fabrication of high-efficiency slab PBG devices with a high yield. To this end, we developed a custom DRIE process that is presented in the following section.

6.3.3   TIME-MULTIPLEXED ETCHING

HAR anisotropic etching technology satisfies the demands of the microelectromechanical systems (MEMS) industry by using sophisticated etching systems to realize structures with feature sizes of up to hundreds of microns [68–70]. However, to adapt these systems to submicron two-dimensional PhC fabrication, there are a number of issues that must be addressed, such as anisotropy, selectivity, aspect ratio, feature-shape-dependent etching and sidewall quality.

In recent work, several researchers have attempted to address these issues. Tokushima et al. [71] demonstrated 1.55 μm light propagation through triangular lattice two-dimensional PhC line-defect waveguide fabricated using a silicon dioxide mask and an electroncyclotron resonance (ECR) plasma etcher with a gas mixture of Cl_2 and O_2. Lončar et al. reported the fabrication of silicon PhC optical waveguides using a chemically assisted ion-beam etching (CAIBE) system with a 1250 V Ar^+

beam assisted by XeF_2 as the reactive gas. Baba et al. [38] reported the fabrication of PhC waveguides using Ni/Ti as etch masks and an inductively coupled plasma (ICP) etching system.

The methods presented in each of these papers rely on using either sophisticated etching systems or additional process steps for the fabrication of PhC devices. For example, high-end etching systems from vendors like Oxford Plasma Technology, rely on the source gases being broken down in a high-density plasma region before reaching the wafer, which is maintained at a small, but well-controlled, voltage drop below the plasma. In less sophisticated RIE systems, this potential is difficult to maintain because it is difficult to achieve a balance between ions and free-radical species generated. Furthermore, there are a number of significant features in the equipment used for DRIE processing that are critical for achieving reliable etches. These include fast pumping, fast-response mass-flow controllers, inductive coupling of power, and heated chamber and pump lines. Also, fast-response mass-flow controllers are required to keep up with the process demands that can include step times as short as 3 s. These added component requirements typically preclude the use of conventional etch systems for DRIE. The development of processes to fabricate PhC devices using conventional etch tools that are available in most research institutions would be a breakthrough for the scientific community. This process could potentially fuel their mass development which would fulfill their promise in revolutionizing optoelectronic integrated circuits.

The etch process developed was carried out in a capacitively coupled reactive-ion etcher (CC-RIE), PlasmaTherm 790 series machine equipped with an 8 in. water-cooled aluminum substrate holder. The chamber was evacuated by a TMP-151C turbo-pump backed by a Leybold D25BCS mechanical pump to pressures around 0.1 mTorr. The etch tool used was supplied with the following process gases: sulfur-hexafluoride (SF_6), tetrafluoromethane (CF_4), helium, argon, oxygen, and hydrogen, which were regulated through mass-flow controllers. The custom etch process was developed for a 3 in. wafer. The 3 in. Si wafer was loaded into the chamber and carbon paint was applied to four corners of the wafer to achieve adequate thermal conductivity between the carrier and the chuck and to fix the wafer in place. Silicon test samples were attached to the 3 in. wafer using vacuum grease.

To achieve deep RIE, three etch characteristics must be obtained, namely

1. Increased etch rate at the bottom of exposed silicon
2. Decreased etch rate at the sidewalls (typically achieved by polymer passivation)
3. Removal of the polymer passivants from the bottoms of the trenches

Since ions are accelerated to the radio-frequency-driven electrodes by the applied bias, the velocity distribution of the ions is anisotropic, with most of the ions moving perpendicular to the device to be etched. This ion directionality leads to increased etch rates at the bottom of the trench compared to that of the sidewalls. The ratio of spontaneous thermal etching (which assists in isotropic etching) to anisotropic etching (due to the ion-enhanced chemical reactions in the plasma) can be controlled by adjusting the ratio of the fluxes of ions to radicals onto the surface of the sample [70].

The addition of noble gases, e.g., helium, to the plasma increases the ion-enhanced chemical reaction rate at the bottom of the trench, leading to enhanced vertical profiles [72]. With the addition of gases like CHF_3 or C_4F_8 (passivant precursors) to the plasma, a sidewall passivation layer can be deposited, which blocks the etching of the sidewalls [73]. Cho et al. [74] have proposed models describing the etching of silicon in carbon–fluorine plasmas assuming that the deposition of CF_x species on the sidewalls forms a layer of $Si\ [(CF_x)]_2$ polymer which blocks the etch reaction, thereby passivating the sidewalls and promoting a vertical profile. However, the etch rate in the vertical direction also decreases with the increasing concentration of the passivant, as the passivation layer deposited at the bottom of the trench has to be constantly removed by ion bombardment in order to allow the chemically active fluorine radicals to etch the trench bottom.

In conventional etching techniques, all the gases are introduced into the plasma as a continuous gas mixture, which means that the deposition of passivation layer, its subsequent removal, and etching are carried out simultaneously. This, however, often leads to insufficient sidewall passivation or to the formation of "grass," (also called black silicon) due to micromasking. To alleviate these difficulties, in my time-multiplexed method, we separated the processes of sidewall passivation, trench-bottom passivant removal, and silicon etching. To this end, the process was divided into cycles, wherein each cycle consists of three separate steps: the deposition of sidewall-passivating polymer film using CF_4 and H_2 gases, and two etching steps using SF_6 and He gases. This method is similar to what is commonly known as the Bosch process [73] and is schematically presented in Figure 6.11.

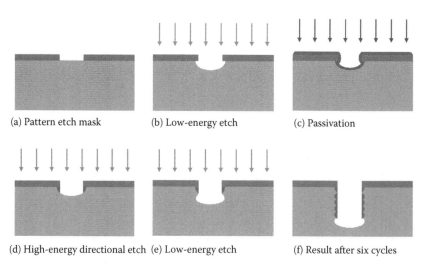

(a) Pattern etch mask   (b) Low-energy etch   (c) Passivation

(d) High-energy directional etch (e) Low-energy etch   (f) Result after six cycles

FIGURE 6.11 Schematic of the custom etch process. (a) Pattern and develop PMMA layer. (b) High-pressure, low-energy isotropic etch. (c) Deposition of CHF_3 passivation layer. (d) High-energy, low-pressure, anisotropic etch via ion bombardment. (e) Repeat the high-pressure, low-energy isotropic etch, and cycle through steps (b)–(d). (f) Etch profile after several repetitions of steps (b)–(d).

By alternating between the deposition and etching steps, anisotropically etched trenches with a high selectivity to photoresist (10:1) were produced. This process enables defining high-quality PhC structures within SOI substrates, as shown in Figure 6.12. Using this HAR custom etch process, high-fill-factor PhC devices ($r/a \sim 0.45$) were fabricated wherein the thickness of the dielectric walls was on the order of 40 nm and etch depths of several microns were achieved.

Figure 6.13 depicts the sidewall profile (surface roughness) of the etched device. Note the ripples can be clearly observed, wherein each ripple represents one cycle. Surface roughness on the order of 20 nm ($\lambda/20$) was measured, which is unlikely to influence the optical properties of PhC devices operating at the wavelength of operation (1550 nm) [75].

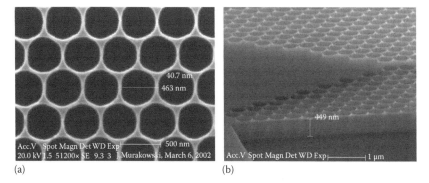

(a)   (b)

FIGURE 6.12   SEM images of (a) top–down view of a high-fill-factor triangular PhC lattice after the custom etch process. Note the minimum thickness of Si separating the air cylinders is on the order of 40 nm. (b) Cross section of the etch profile depicting 10:1 aspect ratio in the sub-100 nm regime.

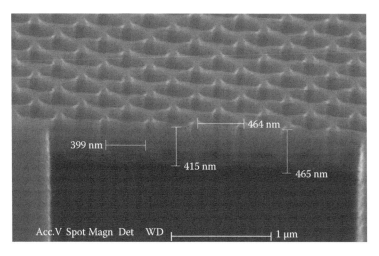

FIGURE 6.13   Zoom-in view of the cross section of the etch profile. Note the formation of ripples due to individual etch and passivation steps.

Following the custom etch process; the next step involves removing the PMMA e-beam resist, as depicted in Figure 6.8e. This can be achieved in two ways: by removal in oxygen plasma, a process commonly referred to as "ashing," or by rinsing the device in acetone followed by methanol and isopropanol. The first method is preferable, as it leaves no residue. Ashing was performed under process conditions of 20 sccm oxygen, 200 mTorr pressure at 100 W RF power for 3 min to completely strip ~200 nm of PMMA.

The final step to complete the fabrication of PhC devices on SOI substrates, as depicted in Figure 6.8f, involves underetching the SiO_2 layer. This is achieved by immersing the SOI substrate in a buffered oxide etch solution (hydrofluoric acid) for 20 min and subsequently rinsing in DI water to leave the Si waveguide and PhC membrane suspended, as shown in Figure 6.14. By generating a suspended membrane, the refractive-index contrast increases in the vertical direction which results in the improvement of the vertical confinement of light within the slab. Furthermore, as opposed to symmetric waveguides, asymmetric waveguides lead to radiation losses since the confined modes can no longer be considered as pure TE- or TM-modes, but rather a superposition of both modes to create TE- or TM-like modes such that one mode contains the dominant field components. A detailed study of the loss mechanisms in PhC line-defect waveguides is presented in Ref. [76].

While the development of the above custom etch process in a CC-RIE proves extremely beneficial for the exploration of planar PhC devices from the point of view of research and development, they are not commercially viable for several reasons which will be described in the following sections.

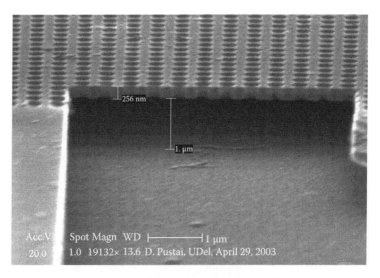

FIGURE 6.14 SEM micrograph of a suspended PhC lattice in silicon, as illustrated in Figure 6.8f. The removal of the 1 μm underlying SiO_2 layer is achieved by a hydrofluoric acid wet etch.

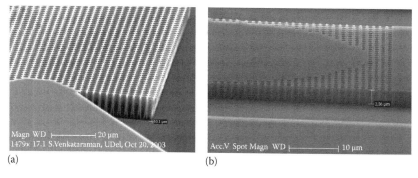

FIGURE 6.15   SEM images of (a) fabricated dispersion engineered THz PhC waveguide on a 10 μm SOI substrate and (b) fabricated THz PhC coupling lens on a 2.56 μm SOI substrate.

6.3.4   Etch Process Toolbox for Advanced Silicon Microshaping

Although the TM-DRIE process was primarily aimed at the fabrication of two-dimensional PhC devices for applications in chip-scale optical interconnects, it has been successfully adapted to several other applications as well. In this section, we present various functional devices that have been realized based on time-multiplexed, fluorine-based ICP-etch processes for advanced silicon micro- and nanoshaping.

6.3.4.1   Terahertz PhC Devices

The terahertz (THz) region of the spectrum, which lies between visible light waves and microwaves, has attracted interest recently because of its unique applications in medical imaging, space communication, and military. The TM-DRIE process has been leveraged in the fabrication of THz waveguides and coupling lenses based on planar PhCs on SOI substrate, as illustrated in Figure 6.15a and b, respectively [77].

6.3.4.2   HAR Etching

The most critical technology for the development of advanced MEMS or micro-optical electromechanical systems (MOEMS) relates to HAR-etching technology. Even in the microelectronics world, HAR etches find applications, for example transistor trench isolation and trench capacitors. Using TM-DRIE process, aspect ratios greater than 30 have been achieved, as depicted in Figure 6.16.

6.4   DISPERSIVELY ENGINEERED PHCS

While the existence of PBG in PhC structures provides an excellent way to manipulate photons and thus allows for a variety of applications as overviewed above, another aspect of PhCs, namely the highly anisotropic dispersion property, is also very important despite being less well known. This property has been widely utilized to build highly dispersive prisms [78]. In 1999, Kosaka et al. [79] observed another outside-gap phenomenon exhibited by a three-dimensional PhC showing collimated light propagation insensitive to the divergence of the incident beam. This phenomenon was called self-collimating phenomena. Subsequently, Witzens et al. [80] theoretically investigated self-collimating phenomena in planar PhCs. Later,

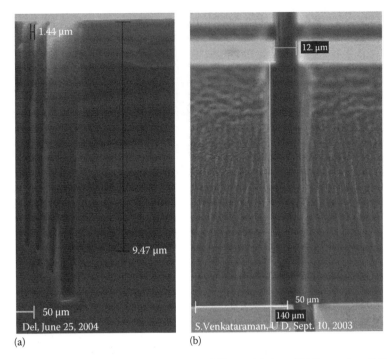

FIGURE 6.16 SEM image illustrating (a) the ability to fabricate HAR trenches (AR > 40) using the TM-DRIE process, and (b) the ability to fabricate deep trenches with etch depths of around 140 μm while still maintaining the anisotropy.

Wu and coworkers [81] experimentally demonstrated self-collimating phenomena in a planar PhC. They also combined self-collimating phenomenon with superprism phenomenon to make a beam deflection device. Later, Chigrin et al. [82] studied self-collimating in two-dimensional PhCs. Inspired by these works, we have extensively investigated silicon PhC outside-gap applications.

Light propagation in a PhC is governed by its dispersion surfaces. Incident light waves propagate in directions normal to the dispersion surface, as shown in Figure 6.17 Curvature of the dispersion surface, as shown in Figure 6.17a, can indicate beam divergence or convergence, whereas the lack of curvature, or straight equifrequency contour (EFC) lines, leads to the so-called self-collimation phenomenon Figure 6.17b.

6.4.1 DISPERSION GUIDING IN PLANAR PhC STRUCTURES

Tailoring the dispersion characteristics of a PhC structure to create unique devices has been an important area of research [82–85]. Self-collimation—also known as autocollimation or natural-guiding—allows a narrow beam to propagate in the PhC without any significant broadening or change in the beam profile, and without relying on a bandgap or engineered defects, such as waveguides. This property can be used for waveguiding and dense routing of optical signals.

To understand self-collimation, consider a planar PhC consisting of a periodic array of cylindrical air holes embedded in a high-index slab. In this case, an

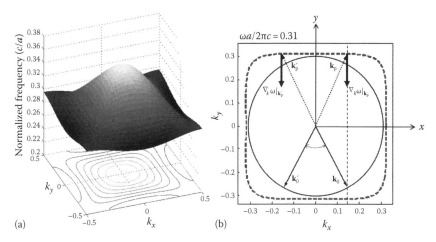

(a)   (b)

FIGURE 6.17 The dispersion surface for a PhC designed to have a square EFC for specified frequencies. (a) A dispersion surface. (b) A square EFC. $\mathbf{k}_0$ is the incident wave vector, $\mathbf{k}$ is the wave vector in the PhC, and $\nabla_k\omega$ is the group velocity in the PhC corresponding to wave vector $\mathbf{k}$.

electromagnetic wave propagating within the plane of the periodic structure interacts with it in both the vertical and lateral directions. In the vertical direction, we only consider field configurations that are confined to the slab by TIR, i.e., those that lie below the light line. On the other hand, in the lateral directions, the interaction is most appropriately interpreted through a dispersion diagram, which characterizes the relationship between the frequency of the wave, ω, and its associated wave vector, $\mathbf{k}$. Dispersion diagrams can be obtained by casting Maxwell's equations into an eigenvalue problem, which can be solved using various computational electromagnetic techniques, such as the PWM [86] or the FDTD method [87]. The set of solutions, takes the form of a dispersion surface, as shown in Figure 6.17a. To obtain such a rendering, one simply computes the eigenfrequencies for wave vectors at all k-points within the irreducible Brillouin zone, and then exploits the appropriate symmetry operations to obtain the entire surface shown in Figure 6.17a.

In general, the dispersion surfaces, obtained by primarily employing the PWM, correspond to index ellipsoids in conventional crystalline optics, where the length from the surface to the Γ-point ($k_x = 0$, $k_y = 0$) is related to the refractive index. In the case of PhCs, however, the dispersion surfaces can take a variety of shapes depending on the lattice type, pitch, fill-factor, or index of refraction of the constituent materials, in addition to the strictly ellipsoidal surfaces of conventional materials. At the same time, we are particularly interested in the EFCs, as they characterize the relationship between all allowed wave vectors in the structure and their corresponding frequencies. For example, while the EFCs of an unpatterned homogeneous silicon slab are circular as depicted by the solid line in Figure 6.17b, the EFCs of a silicon slab with periodic patterns can exhibit a variety of shapes [85]. By carefully selecting the frequency, one can obtain the square shape EFC shown by the dashed contour in Figure 6.17b.

The ability to shape the EFCs, and thereby engineer the dispersion properties of the PhC, opens up a new paradigm for the design and function of optical devices. The importance of the EFC shape stems from the relation

$$\mathbf{v}_g = \nabla_k \omega(\mathbf{k}), \tag{6.1}$$

which says that the group velocity, $\mathbf{v}_g$, or the direction of light propagation, coincides with the direction of the steepest ascent of the dispersion surface and is perpendicular to the EFC, as indicated in Figure 6.17b. In the case of a circular contour, an effective refractive index can be calculated from the radius of the EFC following Snell's law. However, for self-collimation, we desire a square EFC, in which case the wave is only allowed to propagate along directions normal to the sides of the square. As a result, it is possible to vary the incident wave vector over a wide range of angles and yet maintain a narrow range of propagating angles within the PhC.

As with defect-based devices, confinement to the slab is governed by the TIR condition imposed at the core–cladding interface. However, since we are expanding the dispersion diagram by calculating the full dispersion surface in order to obtain the EFCs, we must likewise expand the light line to that of a light cone as shown in Figure 6.18. In this case, if the entire EFC is at a frequency below the light cone, light will remain confined to the slab. In Figure 6.18a we plot the dispersion

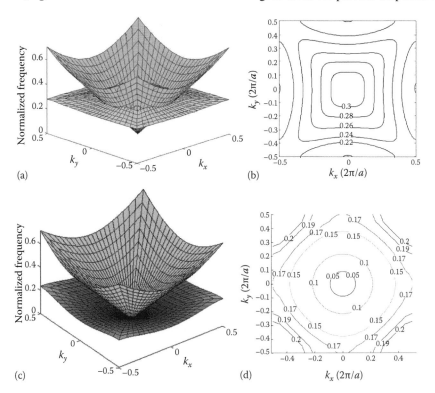

FIGURE 6.18 Nearly planar dispersion surfaces and the light cones for the (a) conduction and (c) valence bands of a PhC and their respective EFCs in (b) and (d).

surface and light cone for the second (conduction) band and find that they intersect around an approximate normalized frequency of $a/\lambda = 0.31$. Therefore, above this normalized frequency light would not remain confined to the slab and couple to the air mode. However, if we operate in the first (valence) band of the PhC, we observe from Figure 6.18b that the entire dispersion surface for this band lies entirely below the light cone. Therefore, when operating in the valence band, the light remains confined to the slab regardless of its frequency, and thus, it is obviously preferable to design a PhC to achieve self-collimation in the first band. However, while feasible, such structures are more difficult to realize for optical wavelengths because the PhC lattice constant and hole radius are much necessarily smaller. Therefore, a majority of the devices presented in this chapter operates in the conduction band of the PhC.

To illustrate the self-collimation phenomenon, we simulated a point source located inside two structures with different EFCs: a homogeneous silicon slab, and one perforated by a square lattice of air holes, with $r/a = 0.3$. For the homogeneous material, the EFC for this material is a circle, and therefore light waves emanate from the source and propagate isotropically within the plane as shown in Figure 6.19a. On the other hand, if the EFC is nearly square, wave propagation due to a point source located at the center of the PhC lattice, as in Figure 6.19b, is limited to the x- and y-directions. While self-collimation can be clearly observed theoretically by introducing a point source into the center of a PhC lattice, this is difficult to achieve in the optical regime except by embedding a source in a PhC consisting of an active material. Therefore, in order to experimentally observe the self-collimation phenomenon in a PhC lattice fabricated in silicon, one must introduce the source in such a way that lateral confinement of the light can be observed.

To characterize the loss in such waveguides, we couple light, of wavelength $\lambda = 1480\,nm$, into the PhC lattice via the input J-coupler where some of the light is scattered due to the impedance mismatch at the silicon/PhC interface as shown in

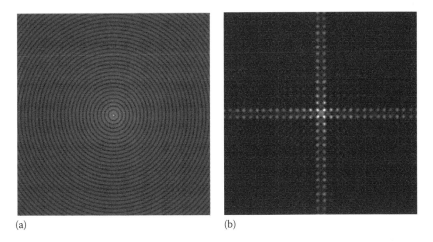

(a)   (b)

FIGURE 6.19   Light propagation from a point source in a (a) homogeneous silicon slab and (b) a silicon slab perforated by a PhC consisting of a rectangular lattice of air holes, wherein $r/a = 0.3$.

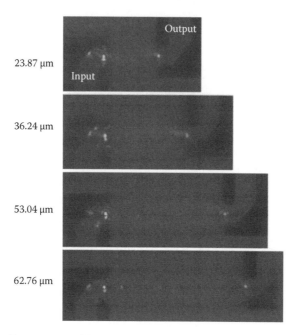

FIGURE 6.20 Image captured by a near-infrared camera of the scattered light, where $\lambda =$ 1480 nm, at the PhC/silicon boundaries. The point located at the output shows how the light is confined laterally within the PhC lattice.

Figure 6.20. We observe another scattered point of light at the opposite end of the PhC lattice, which demonstrates the lateral confinement and self-collimation of the initially divergent light as it propagates along the length of the PhC. The conspicuous absence of a light trail in Figure 6.20 suggests low out-of-plane losses in this guiding structure. In order to quantitatively characterize propagation loss, we fabricated multiple PhC dispersion waveguides, with lengths ranging from 10 to 80 μm, on SOI wafers and employed the cutback method [9]. The loss was obtained from a linear fit of $\log(P_{out}/P_{in})$ vs. waveguide length [88], where P_{in} (P_{out}) is the scattered light at the beginning (end) of the PhC dispersion structure. In using this measure, we assume that the scattered light from each interface is proportional to the amount of light entering and exiting the PhC lattice. From the resulting measurements, propagation loss as low as 1.1 dB/mm is observed which is an improvement over experimental loss measurements for PhC line-defect waveguides [89,90]. Moreover, from three-dimensional FDTD simulations, we find that self-collimating PhCs can, in fact, achieve light propagation with no loss because the wave is guided via the dispersion relation as opposed to the PBG guiding that occurs in line-defect waveguides. In line-defect waveguides, it has been shown that the degeneracy of the mode that exists at the same frequency above the light line causes inherent loss in planar PhCs [7,8]. Additionally, structural deviations along the length of line-defect waveguides due to fabrication tolerances give rise to additional losses as this results in a guided mode shift within the PBG. To this end, this result presents the first ever loss measurements in a self-collimating PhC lattice and demonstrates that, in fact, low-loss guiding can be achieved. Finally, from FDTD simulations, we calculate a 5.3% operational frequency bandwidth and experimentally observe a 3.55% frequency bandwidth in which self-collimation occurs for our designed device.

From the above, it should be clear that such a structure could be used to efficiently guide electromagnetic waves within a planar PhC without the use of channel defects or structural waveguides. Furthermore, the lack of waveguide structures allows multiple beams to be directed across one another in a very high-density fashion without imposing limitations due to structural interactions or cross talk. Ultimately, this allows for very small yet high-capacity photonic circuits by alleviating the limitations of structural interactions that commonly introduce additional loss at these crossing points. The underlying oxide layer is left intact in order to lower the light cone such that the once guided frequencies are now radiated, thereby allowing for qualitative observation of light propagation, although propagation losses are incurred.

6.4.2 NEGATIVE REFRACTION

In most materials, a propagating electromagnetic wave commonly follows a right-handed relationship. In other words, $E \times H = k$ or $S \cdot k > 0$, where S is the Poynting vector, indicating the direction of energy propagation. These materials are referred to as right-handed materials. However, in 1968, Veselago [91] theoretically predicted a class of materials in which the propagating electromagnetic wave exhibited a left-handed relationship, i.e., $E \times H = -k$ or $S \cdot k < 0$. Conversely, these materialas are referred to as left-handed materials. From a material characterization point of view, a right-handed material has a positive phase refractive index, or $n_p > 0$, while a left-handed material has a negative phase refractive index, $n_p < 0$. By juxtaposing these two types of materials, an electromagnetic wave propagating from one medium to another will bend on the same side of the interface normal and undergo negative refraction.

In this section, we demonstrate negative refraction with left-handed behavior at optical wavelengths in a PhC and present related experimental results. To this end, we employ a parabolic focusing mirror to achieve angular selectivity in order to experimentally observe this phenomenon.

To realize left-handed behavior, we design a PhC, consisting of air holes in silicon, such that the conservation of wave vectors, determined by the respective EFCs as shown in Figure 6.21, produces a negatively refracted group velocity vector, v_g, inside of the PhC. The EFCs are calculated at a normalized frequency of $a/\lambda = 0.26$ for a PhC consisting of a square lattice of air holes in silicon background where the radius of air holes is $r = 0.3a$. We consider the energy propagation when a plane wave with a wave vector, k_i, is incident on the PhC structure from a uniform region of silicon at an angle of 10°. To satisfy the boundary conditions, i.e., k conservation along the boundary, we draw a line such that it passes through the intersection point of k_i and the silicon EFC and is perpendicular to the boundary between the background silicon material and the PhC structure. The intersection points between this line and the EFC of the PhC structure in the first Brillouin

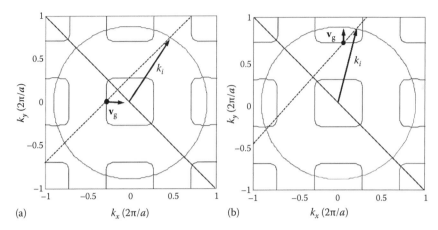

FIGURE 6.21 EFCs of silicon (circular) and the PhC (rectangular) at a normalized frequency of $a/\lambda = 0.26$. Line A represents the boundary between the silicon and PhC region. (a) Negative refraction with left-handed behavior at optical wavelengths. (b) Positive refraction with left-handed behavior at optical wavelengths.

zone would be the end points of the refractive wave vectors if only the condition of k-conservation along the boundary governs the refraction on the boundary. However, in addition to this condition, there are three additional mechanisms that determine the refracted wave vectors and thereby prove that negative refraction is occurring [92]. They are: (1) the refracted wave vectors in an arbitrary material, with a group velocity $\mathbf{v}_g$ in the material, will point away from the source; (2) $\mathbf{v}_g$ will be perpendicular to the EFC of the material and points in the direction governed by $\mathbf{v}_g = \nabla_k \omega(k)$; (3) if the EFC of the material moves outwards with increasing frequency, then $\mathbf{v}_g \cdot k_t > 0$; if it moves inwards, then $\mathbf{v}_g \cdot k_t < 0$, where k_t is the refracted wave vector. The $\mathbf{v}_g$ vector shown in Figure 6.21a satisfies conditions (1) and (2). Since the dispersion surface of the PhC is a downward cone because we are operating in the conduction band, its EFC in the first Brillouin zone moves inwards with increasing frequency, i.e., from condition (3), $\mathbf{v}_g \cdot k_t < 0$. Because $\mathbf{v}_g$ is on the same side of the normal to the air–PhC interface as k_i, negative refraction occurs. Moreover, since $S_t \cdot k_t = \mathbf{v}_g \cdot k_t < 0$, where S_t is the Poynting vector and k_t is the refracted wave vector, the effective phase refractive index is $n_p < 0$ which implies that the material behaves left-handedly [92]. From the EFC, we calculate a negative refractive index of −1.07 for the PhC.

Since the PhC is a periodic structure, wave propagation in the PhC structure follows the form of Bloch modes. In other words, the refracted wave can be expressed as

$$H_z(k_t, r) = h(r)e^{j\vec{k}_t \cdot \vec{r}}, \tag{6.2}$$

where $h(r)$ is a periodic function with a periodicity of the lattice constant a. Therefore, $h(r)$ can be expanded into Fourier series as follows:

$$h(r) = \sum_{n,m} h_{n,m} e^{j\vec{G}_{m,n} \cdot \vec{r}}, \tag{6.3}$$

where $h_{m,n}$ are the Fourier coefficients and $\vec{G}_{m,n}$ are the reciprocal lattice vectors. By substituting Equation 6.3 into Equation 6.2, one can have

$$H_z(k_t, r) = h(r)e^{j\vec{k}_t \cdot \vec{r}} = \sum_{m,n} h_{m,n} e^{j(\vec{k}_t + \vec{G}_{m,n}) \cdot \vec{r}}. \tag{6.4}$$

As such, the refracted wave has many Fourier components with wave vectors $\vec{K}_t + \vec{G}_{m,n}$. From the above analysis, one can see that the incident wave vector k_i is on the same side of the normal of the interface with the energy propagation direction of the refracted wave, indicating negative refraction. Meanwhile, because $S_t \cdot k_t = \mathbf{v}_g \cdot k_t < 0$, where S_t is the Poynting vector of the zeroth-order Fourier component, the effective phase refractive index is $n_p < 0$ as well. Therefore, negative refraction occurs and the material behaves left-handedly.

If the incident angle increases to $20°$, the line drawn perpendicular to the interface does not intersect with the EFC in the first Brillouin zone, rather, it intersects with the EFCs in the second Brillouin zone, as shown in Figure 6.21b. To meet the condition of k-conservation along the boundary between the background silicon material and the PhC structure in addition to the three conditions discussed above, one can have only one refracted wave vector in the second Brillouin zone, such that $\mathbf{v}_g$ points along the k_y direction. In this case, the Poynting vector of the incident wave is on the opposite side of the interface normal with the Poynting vector of the refracted wave. Meanwhile, because in this case $S_t \cdot k_t = \mathbf{v}_g \cdot k_t > 0$, the effective phase refractive index is $n_p > 0$ as well resulting in positive refraction.

To observe this phenomenon, we introduce incident waves at certain angles to the PhC boundary via a structure referred to as a J-coupler [89]. By selectively illuminating sections of the J-coupler through the excitation of higher-order modes in the dielectric waveguide, we can achieve angular selectivity for the wave vectors incident on the PhC. As shown in the steady state field of the FDTD simulation and experimental results as shown in Figure 6.22a, if the right side of the J-coupler is illuminated, we introduce a majority of wave vectors incident on the PhC boundary, cut along one of its $\Gamma - M$ directions, at an angle of $-6°$. Because the PhC behaves left-handedly, a negatively refracted beam is observed within the PhC at an angle of $41.9°$ with respect to the boundary normal. Likewise, if we illuminate the left side of the J-coupler as shown in Figure 6.22b, we generate wave vectors incident on the PhC boundary at an angle of $12.4°$ that produces a negatively refracted wave at $-32.7°$ within the PhC. The angular spectrum of the incident and refracted waves are shown in Figure 6.22c and d for right and left illuminations, respectively. This behavior is observed experimentally as shown in the inlayed images in Figure 6.22a and b. Our PhC structure consists of a square lattice of air holes of $r/a = 0.3$ etched in the 260 nm thick device layer of a SOI wafer.

By controlling the dispersion characteristics of the PhC, we were able to observe negative refraction with left-handed behavior at optical frequencies. Moreover, we

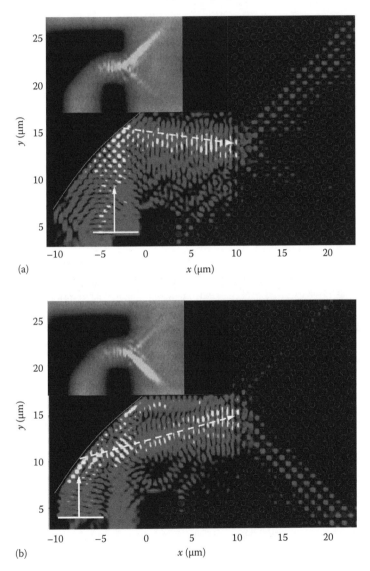

(a)

(b)

FIGURE 6.22 (a) Steady-state Hz field profile of the TE-polarized light as it is incident on the right half of the J-coupler with negative refraction observed in the PhC. Experimental result located in the upper left corner. (b) Steady-state Hz field profile of the TE-polarized light as it is incident on the left half of the J-coupler with negative refraction observed in the PhC. Experimental result located in the upper left corner. (c) Angular spectrum of the incident and refracted field intensity shown in (a), which demonstrates negative refraction in the PhC. (d) Angular spectrum of the incident and refracted field intensity shown in (b).

(*continued*)

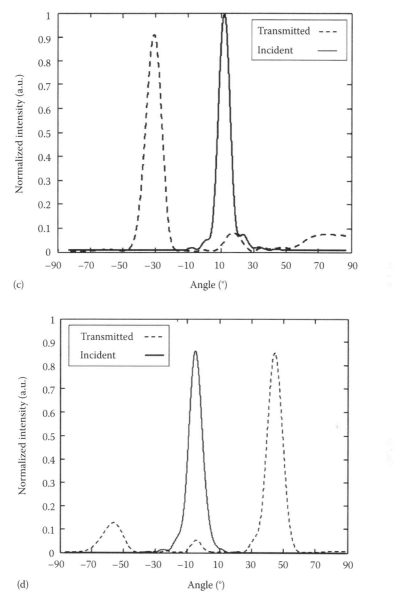

(c)

(d)

FIGURE 6.22 (continued)

experimentally validated negative refraction in a PhC at various incident angles by exploiting the spatial variance of the J-coupler. As a result, a pronounced 74.6° change in the negatively refracted beam was observed by simply traversing the illuminating source across the facet of a 5 μm waveguide. By introducing a spatially sensitive device such as the J-coupler, one can exploit the angular sensitivity of the PhC in order to observe minute changes in the position of the source. Such a device could be used to enhance optical switching and scanning applications.

6.5   FUTURE APPLICATIONS AND CONCLUDING REMARKS

It is clear that PhC technology is just beginning to come of age and many applications yet to be realized lie in wait. Optical components that can permit the miniaturization of an application-specific optical integrated circuit (ASOIC) to a scale comparable to the wavelength of light will be a good candidate for next-generation high-density optical interconnects and integration. In recent years, there has been a growing interest in the realization of PhCs or PBG structures as optical components and circuits. In this paper we applied available computational electromagnetics modeling and simulation techniques to develop and optimize application-specific photonic integrated circuits (ASPIC) in PBG structures for near-infrared or telecommunication applications which will be good candidates for next-generation high-density optical computing systems and interconnects. The implication of this work is the ability to incorporate on-chip optical signal processing and routing, on a scale comparable to the wavelength of light. Currently, optical processing devices tend to have a scale much larger than the wavelength of light, which prohibits their use in "on-chip applications."

Patterning PhC thin films into optical circuits would represent the ultimate limit of optoelectronic miniaturization. Integrated circuits that combine conventional electronics and photonics will extend the integrated circuit revolution into the domain of high-bandwidth optical signals.

Results of this study will hopefully be used to realize a new generation of optoelectronic chips to satisfy the growing demand in terms of next-generation optical telecommunication systems.

PhC telecommunication systems built on a subwavelength scale will not only open many exciting opportunities in integrated optics and high-density optical interconnects, but will also provide the basic building blocks for nanophotonic circuits (NPC) of the future. As the development of semiconductor materials has led to the ongoing electronic revolution, high-density optical interconnects in PhCs may hold the key for achieving the long-sought goal of large-scale integrated photonic circuits (LSPIC) or optical processing.

The prominent contribution of this work includes the ability to incorporate on-chip optical signal processing and routing, on a scale comparable to the wavelength of light. Currently, optical processing devices tend to have a scale much larger than the wavelength of light, which prohibits their use in "on-chip applications." Toward this end, much remains be done to investigate the properties of individual components incorporated in a PhC, and to identify their breakthrough application to be introduced to the telecommunication market, which will indeed benefit from such miniaturization and integration.

REFERENCES

1. A. Adibi, Y. Xu, R. K. Lee, M. Loncar, A. Yariv, and A. Scherer, Role of distributed Bragg reflection in photonic-crystal optical waveguides, *Physical Review B*, 64, art. no.-041102, 2001.
2. E. Yablonovitch, Inhibited spontaneous emission in solid-state physics and electronics, *Physical Review Letters*, 58, 2059–2062, 1987.

3. S. John, Strong localization of photons in certain disordered dielectric superlattices, *Physical Review Letters*, 58, 2486–2489, June 1987.

4. V. Bykov, Spontaneous emission in a periodic structure, *Soviet Physics JETP*, 35, 269–273, 1972.

5. V. Bykov, Spontaneous emission from a medium with a band spectrum, *Soviet Journal of Quantum Electronics*, 4, 861–871, 1975.

6. A. Chutinan, S. John, and O. Toader, Diffractionless flow of light in all-optical microchips, *Physical Review Letters*, 90, 123901, 2003.

7. A. Adibi, R. K. Lee, Y. Xu, A. Yariv, and A. Scherer, Design of photonic crystal optical waveguides with single mode propagation in the photonic bandgap, *Electronics Letters*, 36, 1376–1378, 2000.

8. A. Adibi, Y. Xu, R. K. Lee, A. Yariv, and A. Scherer, Properties of the slab modes in photonic crystal optical waveguides, *Journal of Lightwave Technology*, 18, 1554–1564, 2000.

9. A. Mekis, A. Dodabalapur, R. E. Slusher, and J. Joannopoulos, Two-dimensional photonic crystal couplers for unidirectional light output, *Optics Letters*, 25, 942–944, 2000.

10. E. Centeno and D. Felbacq, Guiding waves with photonic crystals, *Optics Communications*, 160, 57–60, 1999.

11. J. D. Joannopoulos, R. D. Meade, and J. N. Winn, *Photonic Crystals*. Princeton, NJ: Princeton University Press, 1995.

12. C. C. Cheng and A. Scherer, Fabrication of photonic band-gap crystals, *Journal of Vacuum Science & Technology B*, 13, 2696–2700, 1995.

13. M. Loncar, T. Doll, J. Vuckovic, and A. Scherer, Design and fabrication of silicon photonic crystal optical waveguides, *Journal of Lightwave Technology*, 18, 1402–1411, 2000.

14. T. F. Krauss, R. M. DeLaRue, and S. Brand, Two-dimensional photonic-bandgap structures operating at near infrared wavelengths, *Nature*, 383, 699–702, 1996.

15. M. Loncar, D. Nedeljkovic, T. Doll, J. Vuckovic, A. Scherer, and T. P. Pearsall, Waveguiding in Planar Photonic Crystals, *Applied Physics Letters*, 77, 1937–1939, 2000.

16. C. M. Soukoulis, *Photonic Band Gaps and Localization*. New York: Plenum Press, 1993.

17. J. R. Hook and H. E. Hall, *Solid State Physics*. West Sussex, England: John Wiley & Sons Ltd., 1991.

18. M. Plihal and A. A. Maradudin, Photonic band-structure of 2-dimensional systems— The triangular lattice, *Physical Review B-Condensed Matter*, 44, 8565–8571, 1991.

19. P. R. Villeneuve and M. Piche, Photonic band gaps in two-dimensional square lattices: Square and circular rods, *Physical Review B*, 46, 4973–4975, 1992.

20. S. Fan, P. R. Villeneuve, J. D. Joannopoulos, and E. F. Schubert, High extraction efficiency of spontaneous emission from slabs of photonic crystals, *Physical Review Letters*, 78, 3194–3297, 1997.

21. S. G. Johnson, S. H. Fan, P. R. Villeneuve, J. D. Joannopoulos, and L. A. Kolodziejski, Guided modes in photonic crystal slabs, *Physical Review B*, 60, 5751–5758, 1999.

22. K. M. Ho, C. T. Chan, and C. M. Soukoulis, Existence of a photonic gap in periodic dielectric structures, *Physical Review Letters*, 65, 3152–3155, 1990.

23. S. G. Johnson and J. D. Joannopoulos, Block-iterative frequency-domain methods for Maxwell's equations in a planewave basis, *Optics Express*, 8, 173–190, 2001.

24. S. G. Johnson, A. Mekis, J. Fan, and J. D. Joannopoulos, Molding the flow of light, *Computing in Science and Engineering*, 3, 38–47, 2001.

25. C. T. Chan, Q. L. Yu, and K. M. Ho, Order-N spectral method for electromagnetic-waves, *Physical Review B-Condensed Matter*, 51, 16635–16642, 1995.

26. S. S. Xiao and S. L. He, FDTD method for computing the off-plane band structure in a two-dimensional photonic crystal consisting of nearly free-electron metals, *Physica B-Condensed Matter*, 324, 403–408, 2002.

27. J. B. Pendry, Calculating photonic band structure, *Journal of Physics: Condensed Matter*, 8, 1085–1108, 1996.

28. W. Axmann and P. Kuchment, An efficient finite element method for computing spectra of photonic and acoustic band-gap materials—I. Scalar case, *Journal of Computational Physics*, 150, 468–481, 1999.

29. M. Koshiba, Full-vector analysis of photonic crystal fibers using the finite element method, *IEICE Transactions on Electronics*, E85C, 881–888, 2002.

30. W. M. Robertson, Transmission-line matrix modeling of superluminal electromagnetic-pulse tunneling through the forbidden gap in two-dimensional photonic band structures, *Journal of the Optical Society of America B-Optical Physics*, 14, 1066–1073, 1997.

31. D. C. Dobson, J. Gopalakrishnan, and J. E. Pasciak, An efficient method for band structure calculations in 3D photonic crystals, *Journal of Computational Physics*, 161, 668–679, 2000.

32. L. C. Botten, N. A. Nicorovici, R. C. McPhedran, C. M. de Sterke, and A. A. Asatryan, Photonic band structure calculations using scattering matrices, *Physical Review E*, 6404, art. no. 046603, 2001.

33. A. Taflove and S. C. Hagness, *Computational Electrodynamics: The Finite-Difference Time-Domain Method*, 2nd edn. Boston, MA: Artech House, 2000.

34. J. P. Durbano, P. F. Curt, J. R. Humphrey, F. E. Ortiz, and D. W. Prather, FPGA based acceleration of the three-dimensional finite difference time domain method for electromagnetic calculations, in: *Global Signal Processing Expo & Conference (GSPx)*, Palo Alto, CA, 2004.

35. J. P. Durbano, P. F. Curt, J. R. Humphrey, F. E. Ortiz, D. W. Prather, and M. Mirotznik, Hardware acceleration of the 3D finite-difference time-domain method, *IEEE International Symposium on Antennas and Propagation*, 1, 77–80, 2004.

36. J. P. Durbano, P. F. Curt, J. R. Humphrey, F. E. Ortiz, and D. W. Prather, FPGA based acceleration of the 3D finite-difference time domain method, in: *IEEE Symposium on Field Programmable custom computing machines (FCCM)*, Napa, CA, 2004.

37. J. P. Durbano, P. F. Curt, J. R. Humphrey, F. E. Ortiz, D. W. Prather, and M. Mirotznik, Implementation of three-dimensional FPGA based FDTD solvers: An architectural overview, in: *IEEE Symposium on Field Programmable Custom Computing Machines (FCCM)*, Napa, CA, 2003.

38. T. Baba, A. Motegi, T. Iwai, N. Fukaya, Y. Watanabe, and A. Sakai, Light propagation characteristics of straight single-line-defect waveguides in photonic crystal slabs fabricated into a silicon-on-insulator substrate, *IEEE Journal of Quantum Electronics*, 38, 2002, 2002.

39. S. G. Johnson, S. Fan, P. R. Villeneuve, and J. D. Joannopoulos, Guided modes in photonic crystal slabs, *Physical Review B*, 60, 5751–5758, 1999.

40. A. Scherer, O. Painter, J. Vuckovic, M. Loncar, and T. Yoshie, Photonic crystals for confining, guiding, and emitting light, *IEEE Transactions on Nanotechnology*, 1, 4, 2002.

41. J. P. Berenger, A perfectly matched layer for the absorption of electromagnetic waves, *Journal of Computational Physics*, 114, 185–200, 1994.

42. Z. S. Sacks, D. M. Kingsland, R. Lee, and J. Lee, A perfectly matched anisotropic absorber for use as an absorbing boundary condition, *IEEE Transactions on Antennas and Propagation*, AP-43, 14601463, 1995.

43. S. Shi, C. Chen, and D. W. Prather, Plane-wave expansion method for calculating band structure of photonic crystal slabs with perfectly matched layers, *Journal of the Optical Society of America A*, 21, 1769, 2004.

44. P. J. Silverman, The Intel lithography roadmap, *Intel Technology Journal*, 6, 55–61, 2002.

45. M. J. Madou, *Fundamentals of Microfabrication: The Science of Miniaturization*, 2nd edn. Boca Raton, FL: CRC Press, 2001.

46. A. F. Bogenschutz, W. Krusemark, and W. Mussinger, Activation energies in the chemical etching of semiconductors in HNO_3-HF-CH_3COOH, *Journal of Electrochemical Society: Solid State Science*, 114, 970–973, 1967.

47. K. E. Petersen, Silicon as a mechanical material, *Proceedings of the IEEE*, 70, 420–457, 1982.

48. X. P. Wu and W. H. Ko, Compensation corner undercutting in anisotropic etching of (100) silicon, *Sensors and Actuators*, 18, 207–215, 1989.

49. B. Puers and W. Sansen, Compensation structures for convex corner micromachining in silicon, *Sensors and Actuators A*, 23, 1036–1041, 1990.

50. A. Reisman, M. Berkenblit, S. A. Chan, F. B. Kaufman, and D. C. Green, Controlled etching of silicon in catalyzed ethylenediamine–pyrocatechol–water solutions, *Journal of the Electrochemical Society*, 126, 1406–1415, 1979.

51. M. Elwenspoek and H. V. Jansen, *Silicon Micromachining*, Cambridge, U.K.: Cambridge University Press, 1998.

52. S. M. Rossnagel, J. J. Cuomo, and W. D. Westwood, *Handbook of Plasma Processing Technology: Fundamentals, Etching, Deposition, and Surface Interactions*. Park Ridge, IL: Noyes Publications (January 1, 1990), 1990.

53. M. Konuma, *Film Deposition by Plasma Techniques*. Berlin: Springer-Verlag, 1992.

54. D. M. Manos and D. L. Flamm, *Plasma Etching: An Introduction*. San Diego, CA: Academic Press (July 28, 1989), 1989.

55. J. W. Coburn and M. Chen, Optical-emission spectroscopy of reactive plasmas—A method for correlating emission intensities to reactive particle density, *Journal of Applied Physics*, 51, 3134–3136, 1980.

56. H. Kokura, K. Nakamura, I. P. Ghanashev, and H. Sugai, Plasma absorption probe for measuring electron density in an environment soiled with processing plasmas, *Japanese Journal of Applied Physics Part 1-Regular Papers Short Notes & Review Papers*, 38, 5262–5266, 1999.

57. M. Schaepkens, I. Martini, E. A. Sanjuan, X. Li, G. S. Oehrlein, W. L. Perry, and H. M. Anderson, Gas-phase studies in inductively coupled fluorocarbon plasmas, *Journal of Vacuum Science & Technology A-Vacuum Surfaces and Films*, 19, 2946–2957, 2001.

58. K. Miyata, M. Hori, and T. Goto, Infrared diode laser absorption spectroscopy measurements of CF_X ($X = 1$–3) radical densities in electron cyclotron employing C_4F_8, $C2F_6$, CF_4, and CHF_3 gases, *Journal of Vacuum Science & Technology A-Vacuum Surfaces and Films*, 14, 2343–2350, 1996.

59. K. S. Novoselov, D. Jiang, F. Schedin, T. J. Booth, V. V. Khotkevich, S. V. Morozov, and A. K. Geim, Two-dimensional atomic crystals, *Proceedings of the National Academy of Sciences of the United States of America*, 102, 10451–10453, 2005.

60. I. Toyoda, K. Nakamura, and M. Matsuhara, Boundary element analysis of three-dimensional scattering problems by extended integral equation, *Electronics and Communications in Japan, Part 2*, 74, 47–54, 1991.

61. H. Sugai and H. Toyoda, Appearance mass-spectrometry of neutral radicals in radio-frequency plasmas, *Journal of Vacuum Science & Technology A-Vacuum Surfaces and Films*, 10, 1193–1200, 1992.

62. E. Occhiello, F. Garbassi, and J. W. Coburn, Etching of silicon and silicon dioxide by halofluorocarbon plasmas, *Journal of Physics D-Applied Physics*, 22, 983–988, 1989.

63. J. Hopwood, Ion-bombardment energy-distributions in a radio-frequency induction plasma, *Applied Physics Letters*, 62, 940–942, 1993.

64. M. Sekine, Dielectric film etching in semiconductor device manufacturing—Development of SiO_2 etching and the next generation plasma reactor, *Applied Surface Science*, 192, 270–298, 2002.

65. Y. Tanaka, T. Asano, Y. Akahane, B. S. Song, and S. Noda, Theoretical investigation of a two-dimensional photonic crystal slab with truncated cone air holes, *Applied Physics Letters*, 82, 1661–1663, 2003.

66. R. Ferrini, B. Lombardet, B. Wild, R. Houdre, and G. H. Duan, Hole depth- and shape-induced radiation losses in two-dimensional photonic crystals, *Applied Physics Letters*, 82, 1009–1011, 2003.

67. H. Benisty, P. Lalanne, S. Olivier, M. Rattier, C. Weisbuch, C. J. M. Smith, T. F. Krauss, C. Jouanin, and D. Cassagne, Finite-depth and intrinsic losses in vertically etched two-dimensional photonic crystals, *Optical and Quantum Electronics*, 34, 205–215, 2002.

68. S. Aachboun and P. Ranson, Deep anisotropic etching of silicon, *Journal of Vacuum Science & Technology A*, 17, 1999.

69. I. W. Rangelow and H. Loschner, Reactive ion etching for microelectrical mechanical system fabrication, *Journal of Vacuum Science & Technology B*, 13, 2394–2399, 1995.

70. I. W. Rangelow, Dry etching-based silicon micro-machining for MEMS, *Vacuum*, 62, 279–291, 2001.

71. M. Tokushima, H. Kosaka, A. Tomita, and H. Yamada, Lightwave propagation through a 120 degrees sharply bent single-line-defect photonic crystal waveguide, *Applied Physics Letters*, 76, 952–954, 2000.

72. M. Y. Jung, S. S. Choi, J. W. Kim, and D. W. Kim, The influence of He addition on Cl-etching procedure for Si-nanoscale structure fabrication using reactive ion etching system, *Surface Science*, 482, 1119–1124, 2001.

73. R. B. Bosch, U.S. Patents 4855017, 1989; 4784720, 1988.

74. B. O. Cho, S. W. Hwang, I. W. Kim, and S. H. Moon, Expression of the Si etch rate in a CF4 plasma with four internal process variables, *Journal of the Electrochemical Society*, 146, 350–358, 1999.

75. W. Bogaerts, P. Bienstman, and R. Baets, Scattering at sidewall roughness in photonic crystal slabs, *Optics Letters*, 28, 689–691, 2003.

76. J. Murakowski, G. J. Schneider, A. S. Sharkawy, S. Shi, and D. W. Prather, Losses in slab photonic crystals induced by fabrication tolerance, *Proceedings of the SPIE*, 5184, 12, 2003.

77. C. C. Lin, C. H. Chen, A. Sharkawy, G. J. Schneider, S. Venkataraman, and D. W. Prather, Efficient terahertz coupling lens based on planar photonic crystals on silicon on insulator, *Optics Letters*, 30, 1330–1332, 2005.

78. S.-Y. Lin, V. M. Hietala, L. Wang, and E. D. Jones, Highly dispersive photonic band-gap prism, *Optics Letters*, 21, 1771–1773, 1996.

79. H. Kosaka, T. Kawashima, A. Tomita, M. Notomi, T. Tamamura, T. Sato, and S. Kawakami, Self-collimating phenomena in photonic crystals, *Applied Physics Letters*, 74, 1212–1214, 1999.

80. J. Witzens, M. Loncar, and A. Scherer, Self-collimation in planar photonic crystals, *IEEE Journal of Selected Topics in Quantum Electronics*, 8, 1246–1257, 2002.

81. Q. F. Xia, C. Keimel, H. X. Ge, Z. N. Yu, W. Wu, and S. Y. Chou, Ultrafast patterning of nanostructures in polymers using laser assisted nanoimprint lithography, *Applied Physics Letters*, 83, 4417–4419, 2003.

82. D. N. Chigrin, S. Enoch, C. M. S. Torres, and G. Tayeb, Self-guiding in two-dimensional photonic crystals, *Optics Express*, 11, 1203–1211, 2003.

83. H. Kosaka, T. Kawashima, AkihisaTomita, M. Notomi, T. Tamamura, T. Sato, and S. Kawakami, Superprism phenomena in photonic crystals, *Physical Review B*, 58, R10096–R10099, 1998.

84. H. Kosaka, T. Kawashima, A. Tomita, M. Notomi, T. Tamamura, T. Sato, and S. Kawakami, Self-collimating phenomena in photonic crystals, *Applied Physics Letters*, 74, 1212–1214, 1999.

85. J. Witzens, M. Loncar, and A. Scherer, Self-collimation in planar photonic crystals, *IEEE Journal of Selected Topics in Quantum Electronics*, 8, 2002.

86. P. R. Villeneuve and M. Piche, Photonic band gaps in two-dimensional square lattices: Square and circular rods, *Physical Review B*, 46, 4973, 1992.

87. K. S. Kunz and R. J. Luebbers, *The Finite Difference Time Domain Method for Electromagnetics*. Boca Raton, FL: CRC Press, 1993.

88. R. Hunsperger, *Integrated Optics: Theory and Technology*, 4 edn. Berlin, Heidelberg: Springer-Verlag, 1995.

89. T. Baba, A. Motegi, T. Iwai, N. Fukaya, Y. Watanabe, and A. Sakai, Light propagation characteristics of straight single-line-defect waveguides in photonic crystal slabs fabricated into a silicon-on-insulator substrate, *IEEE Journal of Quantum Electronics*, 38, 743–752, 2002.

90. M. Notomi, A. Shinya, K. Yamada, J. Takahashi, C. Takahashi, and I. Yokohama, Structural tuning of guiding modes of line-defect waveguides of silicon-on-insulator photonic crystal slabs, *IEEE Journal of Quantum Electronics*, 38, 736–742, 2002.

91. V. G. Veselago, Properties of materials having simultaneously negative values of dielectric (Xi) and magnetic (Mu) susceptibilities, *Soviet Physics Solid State, USSR*, 8, 3571–3573, 1966.

92. E. Cubukcu, K. Aydin, E. Ozbay, S. Foteinopoulou, and C. M. Soukoulis, Negative refraction by photonic crystals, *Nature*, 423, 604–605, 2003.

7 Fabrication of 3D Photonic Crystals: Molded Tungsten Approach

Paul J. Resnick and Ihab F. El-Kady

CONTENTS

7.1 SYMMETRY, TOPOLOGY, AND PHOTONIC GAPS

Early in the development of photonic crystals, it became evident that the refractive index contrast played a vital role in opening up photonic gaps. A minimum value around ~2.5–3 was found to be the necessary threshold. It was also shown that not any periodic arrangement of dielectric scatterers yields a photonic gap. To date, all the crystal structures that have yielded a full three-dimensional (3D) gap belong to the A7 family of structures [1]. The A7 crystal structures consist of a rhombohedral lattice with a basis of two atoms situated at the crystal positions $\vec{R} = \pm\beta(\vec{a}_1 + \vec{a}_2 + \vec{a}_3)$, where $\vec{a}_1, \vec{a}_2,$ and $\vec{a}_3$ are the primitive lattice vectors defined by

$$\vec{a}_1 = a_0(\varepsilon,1,1), \ \vec{a}_2 = a_0(1,\varepsilon,1), \ \text{and} \ \vec{a}_3 = a_0(1,1,\varepsilon), \ \text{with} \ \varepsilon = 1 - \frac{\sqrt{1 + \cos\alpha - \cos^2\alpha}}{\cos\alpha}$$

where α is the angle between any two primitive lattice vectors. All full 3D gap structures can be produced from this group by proper selection of the parameters, α and β. For example, by choosing $\alpha = 60°$ and $\beta = 1/8$, the diamond structure results as in Figure 7.1. Setting $\alpha = 60°$ and $\beta = 0$, and joining the lattice points by cylinders, the Yablonovite structure results as in Figure 7.2. Similarly, the ISU layer-by-layer structure in Figure 7.3, the spiral rod structure in Figure 7.4, and even the simple cubic structures can be generated by the appropriate choice of parameters. To better understand the rules of thumb for yielding a full 3D band gap, it is imperative to understand how the photonic gap arises. In the next section, we shall follow the argument presented by John et al. [2].

Photonic band gap formation can be understood as a "synergistic interplay" between two distinct resonance scattering mechanisms. On the one hand, there is a microscopic scattering resonance from the dielectric material contained in a single unit cell of the photonic crystal. On the other hand, there is a macroscopic resonance dictated by the geometrical arrangement of the repeating unit cells of the dielectric microstructure.

The microscopic scattering resonance is governed by the local symmetry of the scattering elements. In this case, an incoming light wave is scattered from a one-dimensional (1D) square potential well. Transmission is maximized when the wavelength of the incoming radiation is equal to the width of the well. Reflection is maximized when one-quarter of the wavelength fits in the well. This one-quarter condition is a simple example of the microscopic scattering resonance condition and depends solely on the local configuration of the scattering center.

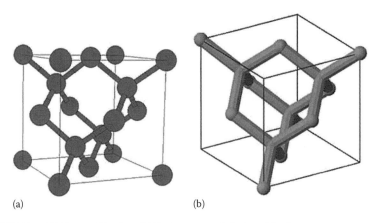

(a)   (b)

FIGURE 7.1   (a) Diamond lattice. (b) Rod connected diamond lattice.

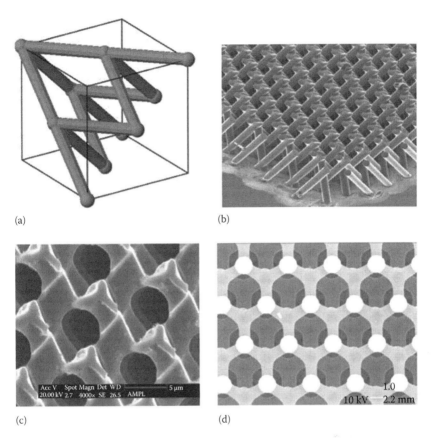

(a)   (b)

(c)   (d)

FIGURE 7.2 (a) A "3-cylinder" structure obtained by collapsing the diamond nodes in the rod-connected diamond structure (Figure 7.1b). (b) SEM of a 3-cylinder structure fabricated with a Sandia National Laboratories LIGA-Electroplating technique. (c) SEM of the "inverse" structure of (b) also known as the "Yablonovite" structure fabricated at Sandia National Laboratories by etching a through a block along 3-axis slanted at 35.26° with respect to the normal to the (111) diamond lattice plane. (d) A planarized version of (c) realized with a microlithographic process using a damascene or "molded" technology at Sandia National Laboratories.

The macroscopic, or Bragg type, of resonance scattering results when there is a periodic arrangement of repeating unit cells of the dielectric microstructure where the spacing between adjacent unit cells is an integer multiple of half of the optical wavelength. A photonic band gap results only if the geometrical parameters of the crystal are such that both the microscopic and macroscopic resonances occur at the same wave length. In addition, both of these scattering mechanisms must be independently quite strong.

The discovery of the A7 family of crystal structures resulted from optimizing the strength of the macroscopic scattering by changing the periodic arrangement of unit cells. The importance of the microscopic scattering was first investigated by Noda et al. [3].

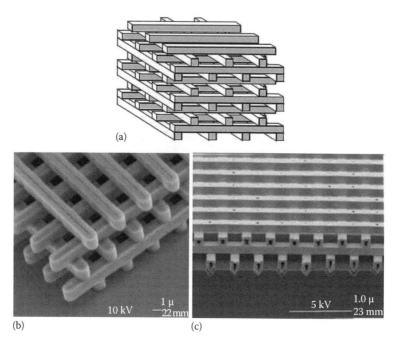

(a)

(b)   (c)

FIGURE 7.3 (a) The ISU layer-by-layer structure consisting of layers of 1D rods with a stacking sequence that repeats after four layers. Within each layer, the rods are arranged in a simple 1D pattern. The rods constituting the next layer are rotated through a 90° angle. The rods in every alternate layer are parallel, but shifted laterally relative to each other by half of a rod spacing. This structure was fabricated at Sandia National Laboratories using a microlithographic damascene or "molded" process technology (b) in silicon and (c) in tungsten.

Noda et al. proposed that a photonic band gap can be opened regardless of the periodic macroscopic arrangement of the scattering centers but is dependent on the local symmetry of those scattering centers. This idea emerged through careful examination of the structures which yielded 3D photonic gaps. They observed that all the structures can be viewed as periodic arrangements of twisted rods [3]. Moreover, they discovered that any periodic arrangements of the twisted rods resulted in a sizable photonic band gap regardless of the underlying symmetry of the lattice. For example, given a face center cubic (FCC) arrangement and fixing the dielectric contrast to 12.25:1, they reported a gap-to-midgap ratio of 17.2% for nontouching dielectric rods in an air background and an even larger gap up to 27.5% was observed when the dielectric rods were allowed to overlap. It is important to point out that even after arranging such twisted rods into FCC, simple cubic, or even body center cubic lattices, the overall symmetry remains that of the A7 family. It is these results that actually motivated the introduction of the tetragonal lattice of square spiral posts suggested by Toader and John [2,4].

The implication of this work is that the "topology" or connectivity of the network does play a vital role in the creation of the gap. Yet, it is not the topology of the individual entities that is of prime importance, rather, that of the high dielectric

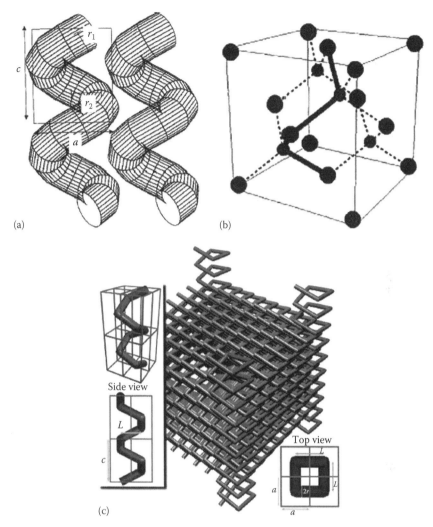

(a)

(b)

(c)

Side view

Top view

FIGURE 7.4 (a) Spiral rods defined in a diamond structure by connecting the lattice points along the (001) crystal direction (b). (From Chutinan, A. and Noda, S., *Phys. Rev. B*, 57, 2006, 1997. With permission.) (c) Tetragonal square spiral photonic crystal. The crystal shown here has a solid filling fraction of 30%. The tetragonal lattice is characterized by lattice constants a and c. The geometry of the square spiral is illustrated in the insets and is characterized by its width, L, cylinder radius, r, and pitch, c. The top left inset shows a single spiral coiling around four unit cells. (From Toader, O. and John, S., *Science*, 292, 5519, 2001. With permission.)

material, especially whether it is in a connected, network topology, or disconnected, cermet topology. This was first pointed out by Ho et al. [5]. As a general rule of thumb, network topology is more favorable for producing large gaps than the cermet topology [6].

The effect of topology on the photonic gaps can be understood by mapping the displacement field intensity to spatially analyze the fields at the top and bottom of the band gap [7]. For simplicity, we shall first consider the case of a two-dimensional square lattice of dielectric cylinders in an air background. In this case, the transverse magnetic (TM) modes have a large band gap and the transverse electric (TE) modes have no band gap. The TE mode electric field vector **E** is in the plane of the crystal and magnetic field vector **H** perpendicular to it. The TM mode has the opposite orientation. Examining the displacement field **D** at the top of the lower band (bottom of the photonic gap) for the TM mode, it is predominantly concentrated in the dielectric rods and little leaks into the air regions. Due to the mutual orthogonality requirement on successive modes, the TM mode resides at the bottom of the upper band (top of the photonic gap) and has a majority of its displacement field concentrated in the air regions. However, from the electromagnetic energy density point of view, the concentration of **D** in the high dielectric yields a lower energy configuration than the case where it is mostly in the air. Therefore, the mode at the bottom of the gap will possess a much lower energy than the one at its top resulting in a large band gap.

For the TE mode case, **E** must remain perpendicular to the rods at all times. Consequently, when the mode at the bottom of the gap tries to concentrate the **D** field in the rods to produce a lower energy configuration, the field penetrates into the air between the cylinders. The mode at the top of the gap, while maintaining its orthogonality to the former mode, is more or less the same and has its entire **D** field in the air regions. The end result is a very small or no energy gap.

Now consider a lattice of air holes in a dielectric host. In this case, the TE modes possess the large gap, while the TM modes have a smaller gap. The TM modes above and below the gap are observed to both concentrate **D** in the dielectric; in the intersections for the lower mode and in the veins in between for the upper mode. Thus, no large gap is produced. The TE modes, on the other hand, confine the lines of **D** to run along the dielectric channels and avoid the air regions. The upper mode, orthogonal to this, forces the **D** field into the air regions, thus opening a large gap.

Extending this analysis to the 3D case explains why a 3D network topology is preferred for large gap production versus a 3D cermet topology. A network topology always has a continuous dielectric path into which the **D** field can concentrate regardless of polarization. The successive mode, which must be orthogonal to this one, will be pushed out of the dielectric and into the air regions, thereby producing a configuration in which two successive modes are different in their energy values opening up a gap. For a cermet topology, this will not be the case. If there is a "low dielectric host" with "high dielectric inclusions," then the boundary conditions on the fields will always force the penetration of the "low dielectric" regions resulting in a reduced gap size or none at all.

7.2 METALLIC PHOTONIC CRYSTALS

Increasing the strength of the micro- and macroscattering resonances implies that the photonic crystal material must have a large refractive index (typically ~3) and negligible absorption (~1 dB/cm). As described above, the individual scattering processes of the material must be independently strong and the material needs to be in a connected, or network, topology. This unique set of requirements has severely restricted the range of dielectric materials that exhibit a photonic band gap. However, these requirements do not limit material selection to dielectrics and metallic materials can be considered. One benefit of metals is the large metallic dielectric function results in a fewer number of periods to achieve a photonic band gap effect [8,9]. However, the inherent metallic absorption, especially at the optical frequency length scales, is problematic. Indeed, most of the proposed metallic or metallo-dielectric photonic crystals have focused on the microwave frequency regions, where the absorption is considerably less [10–16]. However, there are some favorable situations, where the redistribution of the photon wave field, due to the periodicity, prevents the metal from absorbing the light [17]. Under such circumstances, "the light sees the metal sufficiently to be scattered by it, but not enough to be absorbed" [17].

Another consideration in forming metallic photonic crystals is the high conductivity of the metal generating local surface currents changing the intertwined roles of topology and polarization. Kuzmiak et al. [18] studied the case of an array of infinitely long metallic cylinders arranged in square and triangular lattices embedded in vacuum. Their results showed a qualitative difference in the band structures of the two different polarizations.

For the TE modes, they obtained a band structure that was very similar to the free space dispersion except with a number of super-imposed flat bands. For TM modes, no flat bands were observed, rather a finite cutoff frequency below which no propagating modes existed. Because of the orientation of the **E** field, TM modes can couple to longitudinal oscillations of charge along the length of the cylinders, whereas TE modes cannot. Therefore, there exists a cutoff frequency below which there is no propagation due to the vigorous longitudinal oscillations generated by TM-polarized radiation. Because the metal filling fraction is less than 1, these oscillations do not occur at the bulk plasma frequency rather at an effective plasma frequency which scales with the square root of the filling fraction (essentially the square root of the average electron density). Unable to couple to such longitudinal modes, the TE-polarized wave instead excites discrete excitations associated with the isolated cylinders. However, these modes are shifted in frequency and perturbed due to the interactions between the neighboring cylinders. As a result, they appear in the band structure as a number of very flat, almost dispersionless, bands.

For a 3D cermet topology, such as an array of metal spheres, bulk plasma-type oscillations are not possible because the metal is not continuous. Both polarizations show the flat bands caused by the interaction of the modes of the individual spheres [19]. For a 3D network topology, collective oscillations throughout the structure are possible for both polarizations. Consequently, the band structure produces an effective plasma frequency below which propagation is impossible [8].

The first 3D metallic structure was introduced by Sievenpiper et al. [11]. They fabricated a metal wire structure based on a diamond lattice in the centimeter length scale. Here, the structure was created by joining the adjacent lattice points by thick copper wires. In agreement with the above analysis, this network-like structure displayed a forbidden band below a cutoff frequency in the GHz frequency range, as well as a more conventional photonic band gap at a higher frequency resulting from the periodicity of the structure. Defects in the lattice were also introduced, and were observed to allowed modes inside the gap. Almost simultaneously, Ho et al. at Iowa State University (ISU) proposed a structure [12] constructed from layers of a metallic square mesh separated by layers of a dielectric spacer. Their results were in qualitative agreement with the wire diamond lattice, once again identifying a finite cutoff frequency below which no modes could propagate. Defects were also introduced by simply cutting the wires; the result was the appearance of allowed modes below the cutoff frequency.

A theoretical investigation of the behavior of the wire diamond structure was carried out by Pendry [20]. In his calculations, the diameter of the wires was microns rather than the millimeters of the Yablonovitch structure. His results showed the effective plasma frequency of the structure is not only controlled by the average electron density, but also by the inductance of the wires. The net effect was a several orders of magnitude increase in the effective mass of the electrons, consequently reducing the plasma frequency. Unlike the Yablonovitch structure which had a plasma frequency of the same order of magnitude as the lattice spacing, in the micron diameter Pendry structure the square of the plasma frequency was suppressed by a factor of $\ln(a/r)$, where a is the lattice spacing and r is the wire radius. The result is that the plasma frequency is shifted far below the frequency, where diffraction effects occur.

Manufacturable implementations of metallic and metallo-dielectric have been studied by several groups. McIntosh et al. [21,22] proposed the use of an FCC lattice of metallic units embedded in a dielectric background to open up infrared (IR) stop bands. Zhang et al. [23] demonstrated theoretically and experimentally GHz frequency photonic band gaps using dielectric-coated metallic spheres as building blocks. Robust photonic band gaps were found to exist, provided that the filling ratio of the spheres exceeded a certain threshold. However, what was more intriguing about their work was the demonstration of how the photonic band gaps were immune to random disorders in the global structure symmetry. The group also hypothesized that by proper choice of the dielectric spacer and the metal cores, such gaps could be realized in the IR and optical regimes in spite of metallic absorption.

Almost immediately after this, the ISU group [24] investigated theoretically the effects of metallic absorption on the photonic band gap in an all metallic photonic crystal. They argued that metals possessed an IR-to-optical window of frequencies in which metallic absorbance is minimal and can, in fact, be negligibly small with a proper choice of the material. They also showed that by proper choice of the metallic crystal parameters, it is possible to avoid the catastrophic metallic absorption region and overlay the photonic band gap with this preferred window. By satisfying both conditions simultaneously, it was demonstrated that incoming electromagnetic radiation would be rejected by the crystal and negligible absorbance would take place. The effect of intentionally introducing defects was also studied. The group demonstrated that the defect-induced transmission bands suffered from nearly zero absorbance. This opened the door for the possibility of using defects in metallic photonic crystals as IR and possibly optical waveguides.

7.3 MANUFACTURABILITY OF METALLIC STRUCTURES

Here, we consider only metallic structures in the IR regime. For this case, the photonic band gaps created will be insensitive to the type of metal used as all metals in this regime are close to being perfectly conducting. This leaves us with only the effects of the topology and low dielectric host ("the filling") to consider.

As described above, the photonic band gap in cermet topologies is manifested as a band, two band edges, and in network topologies as a frequency cutoff, a single band edge. Therefore, the general rule of thumb is that network topologies will be more resilient to fabrication errors because there is only one fundamental band edge to deal with versus two in the case of cermet topologies. In addition, because the photonic band gap in network topologies essentially extends to infinite wavelengths, the fraction of band gap reduction due to fabrication errors in generally minute and can be overcome by lattice constant rescaling, while in the case of cermet topologies, the band gap shrinkage cannot be compensated in the same manner.

The filler material refers to the low dielectric background in which the metallic entities are embedded. In general, the filler influences the transmission and reflection signatures of the photonic lattice due to the filler's bulk absorption characteristics and has minimal effect on the gap size or location. For the case of a metallic photonic lattice, where the photonic gap becomes quite strong by stacking only a few unit cells, the incoming electromagnetic wave penetration is limited to only a few surface layers in the photonic gap frequency regime, and hence the result of a filler absorption is to reduce the reflectivity of the photonic lattice from near perfect (100%) by a few percentiles (say to ~90%). On the other hand, in the transmission regime of the photonic lattice, where Bloch modes are allowed to propagate, the transmissivity of the lattice will be reduced by a factor proportional to the effective optical path length within the photonic lattice. It is imperative to note two issues at this point: first that electromagnetic radiation propagating though the photonic lattice essentially does so in one path without undergoing multiple scattering since the waves in the transmission regime are propagating Bloch modes of the lattice. Second, the effective optical path length within the photonic lattice is far larger than the simple geometrical length multiplied by the refractive index of the filling, the reason is that at, or close, to the various band edges in the photonic lattice, the actual photonic bands flatten resulting in a reduced group velocity (slow light) increasing the interaction time and the effective absorption cross-section of the lattice. This results in enhanced band edge absorptivity far beyond the typical bulk absorptivity of the lattice constituents.

7.4 FABRICATION OF 3D PHOTONIC CRYSTALS

Various methods have been proposed in the literature for fabricating 3D photonic crystals. The desire for complex, 3D structures that can be mass produced limits the viability of many processing methods. Several researchers have demonstrated a colloidal template method for fabricating synthetic opal structures. A more versatile, layer-by-layer process in which features are defined using microlithography has also been demonstrated for fabrication of highly complex 3D photonic crystals.

7.5   COLLOIDAL TEMPLATE PROCESS

A commonly reported method for fabricating 3D photonic crystals makes use of a colloidal suspension as a template [14]. Through hardsphere-like interactions, a colloidal suspension containing monodisperse, submicron spheres minimizes its free energy by assembling in short range, close packed FCC clusters [15]. The result is the production of random stacks of hexagonal planes, a structure with intrinsic disorder along the c-axis. Charged colloids, on the other hand, yield well-ordered crystals with the FCC arrangement [16]. Such structures have been used to demonstrate the inhibition of spontaneous emission of dye molecules dissolved in the solvent between the spheres [17]. The net negative charge of spheres is counterbalanced by the free ions in the solution. Once these ions are removed, the spheres interact both via long range Van der Waals forces, as well as short range electrostatic repulsion. Under favorable conditions, the colloid undergoes a phase change from a disordered phase to a crystalline FCC structure.

Provided that the monodisperse condition (<5% radial variation) is satisfied within the suspension, a wide range of sphere radii (from 1 nm to 10 μm) can be used to manufacture such crystals. The resulting lattice constants, on the other hand, appear to be governed by the concentration of spheres. The liquid from the colloidal suspension is evaporated, leaving a close-packed opal structure. The interstitial space between particles of the template is filled with the matrix material of choice, such as silicon or silica. Finally, the colloidal particles (e.g., polystyrene latex) are then removed, leaving behind a 3D lattice in the matrix material. Optical properties are determined by the particle size of the colloid template and the index of refraction of the matrix material. This process is depicted in Figure 7.5.

Although long-range order can be achieved in these structures, controlling random defects can be difficult, and excessive random defects can alter the photonic band gap of the material. Furthermore, rather complex processing and limited choices of materials restrict the design space. Finally, the ability to design "defects" into the lattice is highly desirable to create structures such as waveguides or optical cavities. The colloidal template process is not amenable to creating such engineered defects. Because of the limitations to the colloidal template method, no further treatment of this approach will be discussed here.

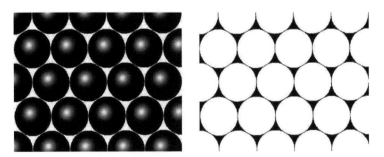

FIGURE 7.5   Colliodal template (left) comprised of ordered, spherical particles; photonic crystal remaining after particles are removed and intercalated material (e.g., silica) is left behind.

7.6   MICROLITHOGRAPHIC PROCESS

Alternative approaches for fabricating 3D photonic crystals have been developed that capitalize on well-established microfabrication processes. A typical foundry for fabrication of microelectromechanical systems (MEMS) or complimentary metal oxide semiconductor (CMOS) devices will possess the capability for film deposition, patterning films using photolithography and film planarization. These unit operations are combined to fabricate devices using a damascene process. A sacrificial film is first deposited on a silicon substrate. Into this film, a pattern is etched; for many applications, this pattern will be a series of lines, the width and pitch of which will be determined by the desired optical properties. Next, the etched features are filled with the structural material of choice, such as polycrystalline silicon. Excess material is removed by chemical mechanical planarization (CMP). This process is then iterated, beginning with the deposition of another sacrificial film, until all of the layers for the design have been deposited. A schematic of this process is depicted in Figure 7.6.

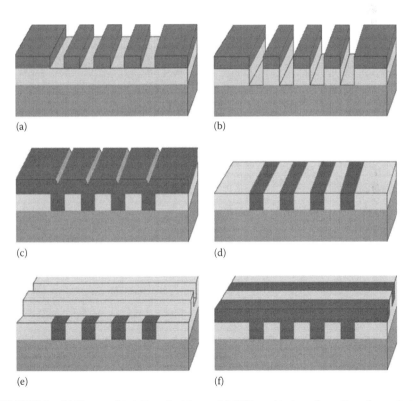

(a)   (b)

(c)   (d)

(e)   (f)

FIGURE 7.6   (a) Photoresist defines first layer; (b) RIE used to transfer pattern from photoresist to SiO$_2$ sacrificial film; (c) conformal tungsten deposition fills features defined in oxide; (d) tungsten CMP used to planarize film; (e) next layer of oxide is deposited and pattern; (f) next layer of structural tungsten is deposited and planarized. Process can be iterated as many times as needed.

A scanning electron micrograph of a simple tungsten ISU-wood-pile structure, fabricated with an iterative damascene process, is shown in Figure 7.7. Note that this method is not limited to rectilinear, or "Manhattan" geometries. Examples of other structures fabricated in a lithographically defined process including square-corner diamond and triangle corner diamond structures are shown in Figure 7.8.

Processes used to fabricate photonic crystal structures leverage heavily off of currently well-established methods used for integrated circuit fabrication. MEMS technologies typically require less stringent design rules that those employed in state-of-the-art integrated circuit fabrication. However, for photonic crystals, the device size scales linearly with the photonic crystal midgap. Thus, for long wavelength (e.g., mid-IR) applications, relatively large lattice spacing is required, which can be easily fabricated with design rule capabilities on the order of $1\,\mu m$. For a crystal with a midgap of $1.5\,\mu m$, the minimum feature size is $0.18\,\mu m$. For visible wavelengths, minimum feature sizes of sub-100 nm will be required, which is in the realm of state-of-the-art fabrication facilities operating at the 90 nm technology

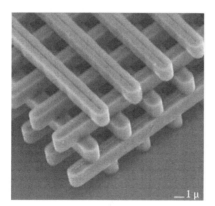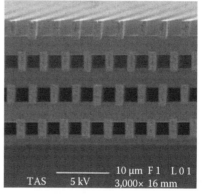

FIGURE 7.7   "Wood Pile" structure fabricated in polysilicon (left) and tungsten (right).

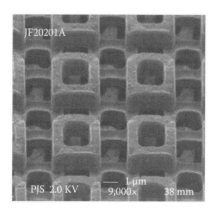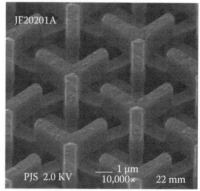

FIGURE 7.8   Square corner and triangle corner diamond structure.

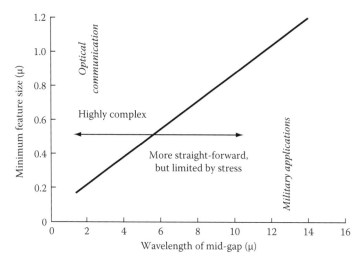

FIGURE 7.9   Minimum feature size versus midgap wavelength.

node or better. Figure 7.9 shows minimum feature size versus the midgap wavelength illustrating the fabrication challenges of shorter wavelength devices.

An important aspect of fabrication with microlithography is the ability to design "defects" into the photonic crystal. Such a feature might be a missing element, and can easily be fabricated by designing the lithographic masks accordingly. The layer-by-layer fabrication method allows these features to be incorporated into any layer, with connectivity from layer to layer. Such "defects" may be used to fabricate optical cavities, waveguides, or other features.

Because the microlithographic process possesses the advantages of flexible design and leverages established processing technology, further discussion of fabrication methods in this chapter will be limited to this damascene or "molded" process. The candidate materials and process considerations for successful integration will be discussed further.

7.7   PHOTONIC CRYSTAL FABRICATION USING A "MOLDED" TECHNOLOGY

To visualize the microlithographic process using a damascene, or "molded" technology, a step-by-step treatment of a typical process may be illustrative. To this end, the fabrication of a woodpile, 3D photonic crystal, will be discussed. For illustration, the structural material will be tungsten, and the sacrificial material will be SiO_2.

The starting material is a silicon wafer, compatible with typical fabrication processes. The process is independent of crystal orientation or doping type and concentration. An anchor film, such as silicon nitride may be used, or the crystal may be anchored directly to the substrate. If the intention is to create a free-floating structure, i.e., one that will be lifted from the substrate after fabrication, then the anchor can be made within the sacrificial film, such that the entire crystal will delaminate from the substrate when the sacrificial film is removed.

The first step is the deposition of the initial sacrificial film; in the example here, this film is a chemical vapor deposition (CVD) SiO_2 film. The thickness of this film will determine the height of the first crystal lattice layer. This thickness will be guided by design considerations, however, from a process standpoint, it is not entirely arbitrary. Etching a very deep, high aspect ratio oxide film may be difficult due to aspect ratio induced etch lag, combined with a relatively low selectivity to the photoresist mask. These process restrictions impose a practical limit to the film thickness. Similarly, because the planarization process has an inherent nonuniformity, as well as a finite selectivity to the oxide material, the amount of remaining oxide may vary significantly across a wafer. If the sacrificial film thickness is small, and selectivity to oxide is low, CMP nonuniformity may result in excessive thinning or punch-through at locations with high removal rates. So, the ability to etch the sacrificial film, and the ability to planarize the structural film with minimal oxide loss must be considered when selecting film thicknesses.

After the first sacrificial film is deposited, the pattern for the first photonic crystal layer is transferred to the film using photolithography. This may be done with a "stepper," a lithographic tool that steps and repeats an image into photoresist across the wafer surface. A mask with the design layout is projected onto the photoresist, typically with a 4× or 5× reduction. When positive photoresist is used (the most common type of photoresist), areas that are exposed to the wavelength of light used by the stepper will dissolve in the developer solution. The result is a replication of the pattern on the original mask, reduced and stepped across the wafer surface, as depicted in Figure 7.10.

The minimum geometry that can be successfully defined is dependent on the resolution of the stepper, and the thickness of the photoresist. Thin photoresist is desirable for patterning fine features; however, sufficient thickness must be present to mask the subsequent etch step.

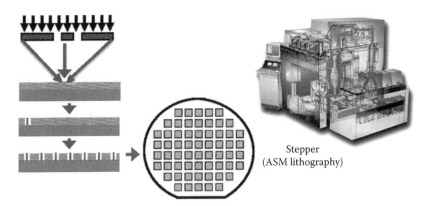

FIGURE 7.10 Illustration of exposing a mask in a stepper tool; stepped pattern across a wafer, and a commercial stepper. (Image of stepper courtesy of ASML Holding N.V.)

After the photoresist pattern is defined, an etch process is performed to transfer the pattern into the sacrificial oxide film. This is most commonly done with a dry, reactive ion etch (RIE) process, which provides a highly anisotropic etch. For oxide etching, a plasma is created using a fluorocarbon gas such as CF_4, C_2F_6, or CHF_3. Ion bombardment activates the etch and provides the anisotropy that is typically desired. Because the first level etch must penetrate into the anchor film, a two step etch that specifically targets the sacrificial film first, and then targets the anchor film may be required. Following the etch, the remaining photoresist mask is stripped, and the wafers are cleaned to remove sidewall polymer that remains after the etch. The process sequence through the first etch is shown in Figure 7.6a.

After the features of the first layer are etched into the sacrificial film, the structural material is deposited into these features (or "molds"). In the example illustrated here, tungsten is deposited using CVD. The precursor molecule for tungsten CVD is WF_6, which may be reduced with silane or hydrogen in the gas phase. Reduction at the silicon surface is also possible for unique applications that require a selective tungsten deposition. The CVD tungsten process provides a highly conformal deposition, capable of filling tortuous features in the oxide film. This CVD process requires a nucleation film to initiate deposition. Because CVD tungsten adheres poorly to oxide films, the nucleation film also serves as an adhesion layer. A thin, sputtered TiN film is used to promote adhesion and nucleation. Although sputtered films typically have poor step coverage, adequate TiN is present for nucleation and adhesion. In order to completely fill the trench, the deposited film must be thick enough such that gap closes off from the edges. In the limiting case of perfect conformality, this means that the required film thickness will be half of the maximum gap spacing. In practice, conformality is not perfect, so the required film thickness will be greater than half of the maximum gap spacing. The practical limits to how thick a film deposition can be will determine the maximum gap spacing. For highly conformal depositions, a reasonable rule of thumb is to limit the maximum gap spacing to the deposited film thickness.

Finally, each layer is completed with CMP, where excess structural material (tungsten) is removed. CMP uses a polishing slurry that selectively removes the target material, with a relatively low attack rate of the sacrificial film. The result is a planar surface, where the sacrificial and structural materials are at the same height. Although the process is selective to the oxide sacrificial layer, the selectivity is finite, resulting in some loss of the sacrificial film. The amount of sacrificial film lost will depend on the amount of overpolishing that will be required to assure complete removal of residual tungsten, as well as the characteristics of the polishing pad and slurry. Furthermore, inherent nonuniformity of the CMP process will create local variation in film thickness. This must be accounted for during the device design phase so that the fabricated device closely matches the as-designed device. Figure 7.6d depicts the process through the first tungsten CMP level.

The entire sequence of sacrificial film deposition, photopatterning, etch, structural film deposition, and CMP is iterated as many times as necessary to complete the design. Each layer is fabricated with a unique mask, such that designed-in "defects" may easily be fabricated.

7.8 FILM STRESS

As deposited, CVD tungsten is typically highly tensile, with a residual stress on the order of 1 GPa. The stress of the sacrificial oxide film depends strongly on the deposition process, and can range from tensile to moderately compressive. The cumulative effect of film stress at each deposition manifests as wafer bow, which can be described by Stoney's equation:

$$\sigma = \frac{Eh_s^2}{(1-v)6Rt_f} \tag{7.1}$$

where
   σ is the film stress
   E is Young's modulus
   v is Poisson's ratio
   h_s is the substrate thickness
   t_f is the film thickness
   R is the radius of curvature

The assumptions in this expression are that both the film and substrate thicknesses are small relative to the lateral dimension; the film thickness is much less than the substrate thickness; and the film and substrate materials are isotropic and homogeneous. For composite films, such as a damascene structure embedded within a sacrificial film, composite values of the physical properties can be used to estimate actual film stress. Although the terminology is counter-intuitive, a compressive film is one that expands against the substrate (cannot be compressed). A diagram of compressive and tensile films is shown in Figure 7.11. By convention, compressive films are assigned a negative stress value (and resulting negative radius of curvature), while tensile films are given positive values.

From a manufacturability standpoint, film stress must be managed carefully, as state-of-the-art semiconductor process tools require very flat substrates for automated handling. Photolithography tools in particular often require very flat substrates, as large numerical aperture optics render the depth of field very shallow. Stoney's equation can be applied to the composite film to determine if the anticipated film stress will result in excessive wafer bow. For example, a lattice structure embedded in sacrificial film that is 3 μm thick and has a net tensile stress of 300 MPa will produce a radius of curvature of about 15 m, assuming the substrate is standard thickness silicon and the initial radius of curvature is 1000 m. The wafer curvature that is acceptable for processing will depend on the toolset that is used, but curvatures less than 15 m can often be troublesome.

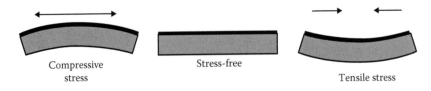

Compressive   Stress-free   Tensile stress
stress

FIGURE 7.11   Effect of film stress on wafer bow.

200 nm   EHT = 10.00 kV WD = 16 mm

FIGURE 7.12   Interlocking beam structure provides additional strength.

Numerous methods exist to manage stress. The design phase should take into consideration the stresses of both the structural film and the sacrificial film, and when appropriate, modify the layout for balanced stress. This may require adding dummy structures or eliminating all nonessential structures, depending on which direction the balance needs to be moved. Although not trivial, the stress of as-deposited tungsten can be manipulated. Oxide films deposited by plasma enhanced CVD may also be tailored for stress by manipulating process conditions.

The consequences of highly stressed structural materials must also be considered. Although the structures may remain sound while embedded in the sacrificial film, upon release, when the sacrificial film is removed, the stress may relieve itself by either buckling the structure (compressive stress) or pulling it apart (tensile stress). This can be mitigated by providing over-etch of the sacrificial film, such that each structural layer notches around the previous layer. The resulting interlocking, or "Lincoln-Log" structure provides additional mechanical resistance to in-plane stress relief. Figure 7.12 shows fabricated interlocking structures. Nevertheless, this interlock results in the reduction of the overall layer thickness only at the cross points, and hence introduces some degree of variation in the layer thickness. Such a behavior has been shown to affect the reflectivity of the lattice, and if severe, can result in the introduction of localized defect modes.

7.9 ALIGNMENT

Photolithographic alignment from one layer to the next will depend largely on the intrinsic capability of the lithographic tools that are used. The amount of alignment error can be less than 10 nm for current state-of-the-art deep ultraviolet tools, to more than 100 nm for older tools often found in MEMS foundries. Scaling errors, exacerbated by the relatively thick, stressy films used for MEMs and photonic crystal fabrication, may also contribute to alignment errors. Although a magnification offset can be used at each photo level to compensate for scaling, a certain amount of error should be expected. Finally, the quality of the alignment marks from the previous layer can affect layer-to-layer alignment. CMP nonuniformity due to pattern density variations may result in varying quality of the previous marks, thus adding some uncertainty to subsequent layer alignment.

7.10 SURFACE ROUGHNESS

The roughness of a printed feature, known in semiconductor fabrication as line edge roughness (LER), must also be considered. This roughness that is transferred from the photoresist mask to the etched film has historically been of little concern to the semiconductor industry because the amount of roughness was quite small compared to the line widths. However, as the technology moves to smaller line widths (<90 nm), even small amounts of roughness, on the order of a few nanometers, contribute significantly to line width variations. Indeed, the International Technology Roadmap for Semiconductors (ITRS) has targeted an LER of 3.4 nm (3σ) for the 65 nm technology node, with smaller LER targeted for subsequent technology nodes (1). For relatively large, long-wavelength optical devices, LER is unlikely to be significant; for shorter wavelength devices, LER may create considerable difficulty. The cause of LER, translated from the photoresist mask, is not well understood, and is further complicated by the challenges of measuring small values of roughness with reasonable precision and accuracy.

7.11 SIDEWALL PROFILE

In addition to roughness, sidewall profile must be considered. Oxide etch typically produces a trench with sidewall taper, such that the top of the trench (and hence the top of the photonic crystal beam) will be wider than the bottom. The extent of the taper can be controlled by modifying the etch parameters (chemistry, ion energy, etc.). Typical values for taper range from nearly 0° off-normal (vertical) to about 5° off-normal. When optimizing taper, a compromise must often be struck between taper and other factors such as etch selectivity to the mask material.

7.12 RELEASE ETCH

After completing the above steps where each layer is defined in a sacrificial oxide film, the sacrificial film must be removed. For SiO_2 films, this is done with a concentrated aqueous HF solution. This "release" etch chemistry must have very high selectivity to the structural material, as it will be exposed to the etch solution. For rigid, noncompliant structures, parts may simply be dried in air after a thorough rinse with water. For devices that are designed to be compliant, capillary force may cause moveable elements to collapse. For these devices, more sophisticated drying methods such as critical-point drying may be required. In critical point drying, supercritical CO_2 is used to displace a miscible solvent such as isopropanol or methanol such that the parts are not exposed to a liquid–vapor interface. Without the liquid–vapor interface, a meniscus does not form, and therefore there is no capillary force to pull structures together.

7.13 MEASUREMENT METHODS, TEST STRUCTURES, AND FAILURE MODES

Because standard integrated circuit fabrication methods are used for the fabrication technology, standard in-line metrology may also be used. Typically, linewidth (critical dimension), film thickness, wafer curvature, and particles are measured. Film thickness is particularly difficult for sacrificial films used in this technology, as the film thickness may be quite large, and measurement areas must be free of damascene tungsten. CMP removal rates are dependent on the underlying feature density; therefore, measurement in open areas may not accurately reflect the film thickness in high density areas. If a thick structure is to be fabricated, the ability to measure thick films by ellipsometry must be considered.

In addition to standard in-line measurements, designers often include test structures for evaluating device integrity or film properties after the release etch has been performed. Such structures may be used as a diagnostic tool to aid subsequent design iterations, or to guide process modifications. For example, in-plane residual stress may be quantified by using a simple bent-beam strain gauge. A SEM micrograph of a bent-beam strain gauge, fabricated in molded tungsten, is shown in Figure 7.13. In this structure, the beam on the right is anchored to the substrate at the center of the beam; the beam on the left is free to move after the sacrificial film is removed.

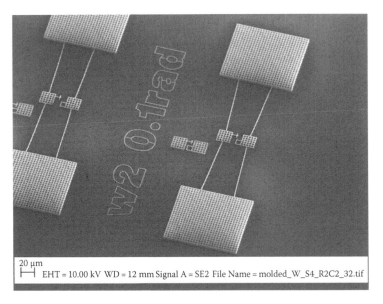

20 µm
⊢—⊣ EHT = 10.00 kV WD = 12 mm Signal A = SE2 File Name = molded_W_S4_R2C2_32.tif

FIGURE 7.13 SEM micrograph of bent beam structure. Rather than using Verniers for measuring displacement, this structure uses a reference cell.

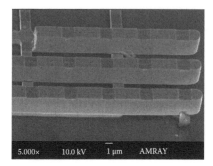 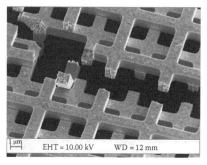

FIGURE 7.14 Failures associated with interlevel adhesion (left) and intralevel stress cracking (right).

If the structural film is under tensile stress, the beam straightens out, thus decreasing the effective length of the beam; for compressively stressed material, the free beam will buckle, thereby increasing the effective beam length.

This strain gauge is interrogated optically; the position of the free beam is measured relative the fixed beam, and the total movement of the beam is determined by comparing the final position to the reference structure. Automated image recognition software may be used for quantifying the amount of beam movement. Analysis of the beam mechanics of these strain gauges can be found in the literature [18,19].

The high residual stress possessed by the tungsten mechanical layers may result in lattice failure once the sacrificial oxide is removed. These failures may include interlayer delamination or within layer stress cracking, as shown in Figure 7.14. Various design modifications, such as using reinforcements or additional contacts to previous layers, may be needed to enhance the robustness of the released structure. Also, the use of compliant structures (springs) to allow in-plane lattice motion can be considered to help relieve residual tungsten stress.

7.14   SUMMARY

Various methods have been proposed for fabricating 3D photonic crystals. Microfabrication using established semiconductor and MEMS processes offers substantial design flexibility while leveraging existing infrastructure. The tungsten/oxide material system has been demonstrated for fabricating highly complex crystals, but other material systems may be employed using similar methodology. Regardless of the material system used, consideration must be given to material compatibility, physical properties, residual stress, and the reliability/durability of the released structure.

ACKNOWLEDGMENTS

Sandia is a multiprogram laboratory operated by Sandia Corporation, a Lockheed Martin Company, for the United States Department of Energy's National Nuclear Security Administration under contract DE-AC04-94AL85000.

REFERENCES

1. S. D. Cheng, R. Biswas, E. Ozbay, S. McCalmont, G. Tuttle, and K. M. Ho, 1995, *Appl. Phys. Lett.* 67, 3399.
2. A. Blanco, E. Chomski, S. Grabtchak, M. Ibisate, S. John, S. W. Leonard, F. L. Meseguer, H. Miguez, J. P. Mondia, G. A. Ozin, O. Toader, and H. M. Van Driel, 2000, *Nature* 405, 437.
3. A. Chutinan and S. Noda, 1997, *Phys. Rev. B* 57, 2006.
4. S. R. Kennedy, M. J. Brett, and O. Toader et al., 2002, *Nano Lett.* 2(1), 59.
5. K. M. Ho, C. T. Chan, and C. M. Soukoulis, in *Photonic Band Gaps and Localization*, NATO ASI Series B, Vol. 308, Ed. C. M. Souklous, Plenum, New York, 1993.
6. E. N. Economou and M. M. Sigalas, 1993, *Phys. Rev. B* 48, 13434.
7. R. D. Meade, A. M. Rappe, K. D. Brommer, and J. D. Joannopoulos, 1993, *J. Opt. Soc. Am. B* 10, 328.
8. M. M. Sigalas, C. T. Chan, K. M. Ho, and C. M. Soukoulis, 1995, *Phys. Rev. B* 52, 11744.
9. S. Fan, P. R. Villeneuve, and J. D. Joannopoulos, 1994, *Phys. Rev. B* 54, 11245.
10. E. R. Brown and O. B. McMahon, 1995, *Appl. Phys. Lett.* 67, 2138.
11. D. F. Sievenpiper, M. E. Sickmiller, and E. Yablonovitch, 1996, *Phys. Rev. Lett.* 76, 2480.
12. J. S. McCalmont, M. Sigalas, G. Tuttle, K. M. Ho, and C. M. Soukoulis, 1996, *Appl. Phys. Lett.* 68, 2759.
13. M. M. Sigalas, G. Tuttle, K. M. Ho, and C. M. Soukoulis, 1996, *Appl. Phys. Lett.* 69, 3797.
14. W. Luck, M. Klier, and H. Wesslau, 1963, *Naturwissenschaften* 50, 485.47.
15. K. Busch and S. John, 1998, *Phys. Rev. E* 58, 3896.
16. J. E. G. Wijnhoven, L. Bechger, and W. L. Vos, 2001, *Chem. Mater.* 13, 4486.
17. J. Martorell and N. M. Lawandy, 1990, *Phys. Rev. Lett.* 65, 1877.
18. L. Lin, A. P. Pisano, and R. T. Howe, 1997, *J. MEMS* 6(4), 313.
19. Y. B. Gianchandani and K. Najafi, 1996, *J. MEMS* 5(1), 52.
20. J. B. Pendry, A. J. Holden, W. J. Stewart, and I. Youngs, 1996, *Phys. Rev. Lett.* 76, 4773–4776.
21. K. A. McIntosh, L. J. Mahoney, K. M. Molvar, O. B. McMahon, S. Verghese, M. Rothschild, and E. R. Brown, 1997, *Appl. Phys. Lett.* 70, 2937–2939.
22. J. A. Oswald, B.-I. Wu, K. A. McIntosh, L. J. Mahoney, and S. Verghese, 2000, *Appl. Phys. Lett.* 77, 2098–2100.
23. W. Y. Zhang, X. Y. Lei, Z. L. Wang, D. G. Zheng, W. Y. Tam, C. T. Chan, and P. Sheng, 2000, *Phys. Rev. Lett.* 84, 2853–2856.
24. I. El-Kady, M. M. Sigalas, R. Biswas, K. M. Ho, and C. M. Soukoulis, 2000, *Phys. Rev. B* 62, 15299–15302.

Index

A

Active matrix liquid crystal display (AMLCD), 46
Analog lithography
  axially symmetric elements
    aspheric lens profile, 104
    circular binary phase-grating profile, 99, 102
    convolution process, 100, 102
    e-beam direct writing technique, 100–101
    immersion technique, 100, 103
    lens parameters, 104
    microlenses fabrication, 101, 103
    rotational symmetry, 99
    scanning electron microscope (SEM) picture, 105–106
    target design profile, 105
  grayscale masking, 83–84
  micro-optics photoresist processing
    analog vortex elements, 99, 101
    resist profile, 98
    vortex micro-optics, 99–100
  phase mask
    adjacent diffraction orders, 84
    amplitude transmittance, 86
    analog intensity transmittance, 87
    analog photoresist sculpting, 88
    analog resist profile formation, 89
    2D binary phase-grating, 90
    2D resist height profile, 94
    duty cycle, 86–87, 93
    fabrication, 97
    far field diffraction field, 86
    fill factor, 89
    finite numerical aperture, 84
    Fourier spectrum, 86
    GCA g-line stepper, 87–88
    grating efficiency, 93
    imaging intensity, 88
    light scattering, 88–89
    microprism, 92–93, 96
    numerical deconvolution method, 94
    partial coherence factor, 85
    π phase depth, 87
    photolithographic stepper, 84–85
    pupil diagram, 84–85
    sixth-order polynomial function, 94
    V-groove structure design, 94–95
    zeroth-order diffraction efficiency, 86–87
    zeroth-order grating transmittance profile, 93–94
  photoresist characteristics
    convolution curve, 92
    SPR 220-7 resist vs. exposure time, 91
    thickness vs. duty cycle, 92–93
Anisotropic etching, 7

C

Chemically assisted ion-beam etcher (CAIBE), 7
Chemical mechanical planarization (CMP)
  microlithographic process, 201
  "molded" technology, 205
Coarse wavelength division multiplexing (CWDM), 53–54

D

deBroglie wavelength, 111
Diffractive optical elements (DOEs)
  nanoimprint lithography (NIL), 144
  self-electro-optic effect device (SEED), 117–119
Dispersively engineered planar photonic crystals
  dispersion guiding
    autocollimation/natural-guiding, 174
    core–cladding interface, 176
    dispersion diagram, 175
    group velocity, 176
    light cone vs. dispersion surface, 176–177
    light propagation, 177
    light scattering, 177–178
    line-defect waveguide, 178–179
    structural waveguide, 179
  negative refraction
    air holes, silicon, 179
    angular spectrum, 181–183
    Bloch mode, 180
    Brillouin zone, 179–180
    electromagnetic wave propagation, 179
    equifrequency contour (EFC), 180
    Fourier series, 181
    J-coupler, 181–183
    refracted wave expression, 180–181
    steady-state Hz field, 181–182
  self-collimating phenomena, 173–174
Dose exposure characterization process, 91